Evolutionary Systems Development

A practical guide to the use of prototyping
within a structured systems methodology

John Crinnion

Department of Business Systems Analysis
City University, London

FINANCIAL TIMES
Prentice Hall

An imprint of **Pearson Education**

Harlow, England · London · New York · Reading, Massachusetts · San Francisco
Toronto · Don Mills, Ontario · Sydney · Tokyo · Singapore · Hong Kong · Seoul
Taipei · Cape Town · Madrid · Mexico City · Amsterdam · Munich · Paris · Milan

Pearson Education Limited
Edinburgh Gate
Harlow
Essex CM20 2JE
England

and Associated Companies throughout the world

Visit us on the World Wide Web at:
http:// www.pearsoneduc.com

First published in Great Britain 1991

© J. Crinnion 1991

ISBN 0 273 03260 7

British Library Cataloguing in Publication Data
A CIP catalogue record for this book can be obtained from the British Library

10 9
04 03 02 01 00

Printed and bound in Great Britain by Biddles Ltd
www.biddles.co.uk

PREFACE

This book is about two apparently conflicting methods of computer systems development, **Structured Systems Analysis & Design**, and **Prototyping**. Its purpose is to illustrate how the best aspects of both can be combined into a single approach, which is both simple and flexible, and at the same time thorough and cost-effective.

I was recently invited to sit in on a one-day seminar given internally by one of the larger government departments. Six different speakers, each representing a project team working in a separate division (and location), described their projects, and their experiences using fourth generation languages and prototyping techniques. Unusually perhaps, all the speakers were capable, informative and entertaining, and they described developments which contrasted strikingly in form and size. In the discussion that followed, there were important lessons to be learnt from each of the applications, and the seminar was undoubtedly a great success. However, one thing was absolutely clear even before the discussion began: **all six of the development teams were working on entirely different interpretations of the term 'prototyping'!**

The book attempts to assemble the lessons from many such encounters, and to establish some form of 'best practice', bearing in mind the range of hardware and systems software within which analyst/prototypers must work.

By contrast, the field of structured systems analysis and design is much more formalised. We live in the 'methodology' era of system development; the tools and techniques for modelling business and computer systems are bundled into integrated sets, each of which is put forward as a general purpose solution to the problem of system development. Although many of these methodologies are now well-established, and the benefits of such an approach are taken for granted, there are some recognised weaknesses, and notable criticisms have been aimed at methodologies, in both their individual and generic form.

One such criticism comes from Greg Grosch, himself one of the developers of the *Stradis* methodology, a long-established and well-proven example of the genre. He points out that systems methodologies 'simplify' the task of development by cutting down the available options, thereby limiting and controlling the variety of possible solutions. The level of flexibility and scaleability that even the best of the structured methodologies provide has long been under question.

A recent Butler Cox report highlighted another aspect of structured systems analysis which gives rise to some concern. The report suggested that structured systems methodologies were not living up to their potential, and that many of the hoped-for benefits were not being recognised in a large percentage of cases. One of the main reasons put forward was that many of the analysts did not properly understand the models they were building or the techniques they were using. Often these analysts were 'going through the motions', carrying out the detailed steps within the methodology, but without appreciating why they were necessary.

One of the objectives of this book is to present the major techniques of structured systems analysis in their most basic form, and to explain the purpose of each aspect or nuance within the technique, so that it can be understood, irrespective of the methodology in which it is being used.

To go back, briefly, to the subject of prototyping, one of the premises of this book is that the majority of business information systems are human-activity systems, and, like the human beings that run them, they are organic and 'evolve', gradually being modified as weaknesses are found and corrected, and potential improvements are applied. Some forms of prototyping support this view of the development process more so than others, and it is the evolutionary form of prototyping which is favoured strongly in this work.

We are living in exciting times from the point of view of advances in systems development techniques. Not only are there decisions to be made concerning structured systems analysis and prototyping; there are new techniques, and even technologies, queuing up for the systems developer's attention. Examples of these include the ever-improving field of CASE, the promising new subject of object-orientation, and the notion of Hypertext and semantic data management. Even the potential of knowledge-based and expert systems, which, for the last twenty years, has alternately excited and frustrated the IT profession, has at last been realised to the extent that business expert systems are now being used regularly, and are proving themselves to be efficient and cost-effective. Never before has there been such a range of options for the analyst to consider; (this is one of the reasons why Grosch's comments are so apposite).

In no approach to systems development can it be guaranteed that all new ideas and techniques in the field can be made to fit within its framework or philosophy. However, it is argued here that the structure of integrated methods and concepts put forward under the name of Evolutionary Systems Development incorporates a level of flexibility which is more likely to allow for future progress and enhancement than any of the other approaches currently available to the Systems Developer.

CONTENTS

Part 1

Part 2

Part 3

ACKNOWLEDGEMENT

I wish to express my gratitude to two of my colleagues, Alan Boadle and Chris Browne, who worked conscientiously through a less-than-perfect draft of this book, and provided comments and suggestions which have led to major enhancements to the text.

Credit must also be given to five 'generations' of students from the DBDP course at the City University. Many insights have come from their reactions and responses, and their commitment and sheer hard work have made a substantial contribution to the evolution of the Systemscraft methodology.

Having said that, none of the above-mentioned can take responsibility for any of the errors or weaknesses which the reader may find in this book; that credit is all mine!

For Marsley

Part 1

Part 1

1 INTRODUCTION AND OVERVIEW

One of the most pressing problems facing systems development managers and senior systems analysts in modern organisations is how to provide the full range of methods and facilities necessary for the analysis, design and construction of all kinds of business computer systems. The problem stems from the wide variety of size and type of such systems. There are, for example, within an average sized organisation:

1. Large and complex systems requiring many person-years of development, and small systems which can be designed and developed within a matter of days,

2. Systems at the operations level, where specific processes must be clearly defined, and other systems at the managerial and policy level, where the requirement is to make information available in the form most convenient for ad hoc processing,

3. Systems which require a 'real-time' interaction between the computer and the environment, other systems where users will conduct an on-line dialogue with the computer, and yet further systems where a traditional 'batch processing' approach is more appropriate,

4. Systems of a fixed and permanent nature, which can be completely defined once and will remain stable, and others (the vast majority) which must be allowed to change and grow with the natural evolution of the company,

5. Systems which interact with and make use of converging technologies such as data communications and office automation.

What is required is a complete set of methods, standards and guidelines for systems development, which can be introduced and enforced within the organisation, thereby providing the means for control and continuity, and promoting stability within the systems development process. Such an approach is nowadays referred to as a 'methodology'.

Paradoxically, any such approach must also be flexible enough to allow the incorporation of new ideas and techniques. As a result, even the most well established of the methodologies currently on the market (eg. SSADM, Information Engineering, etc.) are constantly being reviewed and adjusted.

Over the last ten years, the dominant approach to systems development has been that of 'structured systems analysis and design', and most of the larger business organisations have incorporated one of the methodologies associated with the approach. These include several generations of the Yourdon methodology, Gane and Sarson's Stradis, BIS Modus, Jackson Structured Design, and a number of different variants of both SSADM and Information Engineering.

More recently, there has been a gradual acceptance among systems development departments of a technique known as 'prototyping'. The use of the technique is dependent upon the existence of what are known as 'fourth generation environments'; packages of systems development software including very high level programming languages. Together, the new techniques and the environments have proved their worth by providing such benefits as earlier delivery, reduced development costs and more user-acceptable products.

One of the most powerful forms of prototyping involves the gradual 'evolution' of the system by means of a series of additions and adjustments to the original prototype, the final version of the prototype becoming the live system. This approach is known as Evolutionary Systems Development.

Although many of the structured systems methodologies mentioned earlier have attempted to incorporate the prototyping technique within their set of methods and standards, it has not generally been possible to include the 'evolutionary' concept of prototyping within their frameworks, which are based on the traditional structure of the systems development life cycle. This apparent clash of philosophies has placed some organisations in the position of having to choose between the benefits and dangers of two conflicting approaches.

It is the express purpose of this book to illustrate that the undoubted benefits which can be derived from evolutionary development do not have to be achieved at the expense of the advantages gained from the use of a rigorous structured systems methodology. The apparent conflict can be resolved, providing an approach which incorporates the best of both worlds.

There are now a number of well established evolutionary development methodologies in operation in both Britain and the USA, and the book will describe in general this new 'class' of methodology, quoting specific examples where applicable. However, in order to illustrate the full range and potential of such a methodology, one particular evolutionary methodology is covered in great detail. This methodology, known as 'Systemscraft', was developed at the City University, London, and is now being used in several major organisations.

Unlike many of the earlier generation of structured systems methodologies, Systemscraft was not designed as a rigid 'cookbook' approach to the development process. It is now generally recognised that a good methodology should be flexible enough to be adjustable to suit all kinds of environment and situation; in particular it should be possible to 'tailor' it at three different levels:

1. Each organisation must be able to make the methodology fit the company style and ethos,

2. Within the organisation, it may be necessary to have several slightly different versions of the design part of the methodology, each geared to a particular hardware/software environment,

3. For each individual project, the methodology must be 'tuned' to address the specific circumstances. This may involve the inclusion or exclusion of particular techniques, or adjustments in the weighting of some of the stages.

So the Systemscraft methodology is put forward in this book as a working example, a 'prototype' from which any systems development organisation, large or small, can evolve its own in-house Evolutionary Development Methodology!

The final chapters of the book again address the broader topic of evolutionary development, and examine some of the most important implications. These include, for example, the major change in relationship between the analyst/designer and the owner/user of the system, the changes in approach to the feasibility study, and the ramifications for project management.

This last point is particularly important, as many of the existing prototyping methodologies have failed to provide enough guidance on the subject of project management, and have therefore left the impression that the prototyping technique is only of use in small developments. This is patently not true. In fact, it is easier to delineate work units in a large project when using an evolutionary methodology, because the product of each stage (a prototype) is a concrete piece of work recognisable by both user and analyst alike, whereas the product of many of the stages in a normal structured systems development is an abstract graphical model. The justification for the use of an evolutionary methodology very much hangs on the proof that such an approach can be effectively project managed, and attempts are made throughout the book to provide the necessary evidence.

Another aspect of evolutionary development which is stressed in the final chapters of the book is that of flexibility. There are many important new and developing ideas which, within the next few years, will again cause us to review our approach to systems development. For example, the continuing progress of CASE tools, the assumed potential of object-oriented design, and the eventual arrival of knowledge-based systems methods, could all lead to changes in the way business computer systems are developed. Whether existing methodologies can themselves 'evolve' to incorporate these changes depends on their level of flexibility, and on the basic philosophy which they espouse. For instance, at the moment it is generally believed that knowledge-based systems in particular can only be developed using some form of evolutionary approach, so methodologies based strictly on the traditional systems development life-cycle may not lend themselves easily to such adaptation.

The remainder of this introductory chapter is concerned with defining the major topics of the book: structured systems analysis and design, and prototyping.

1   STRUCTURED SYSTEMS ANALYSIS

Structured systems analysis is a generic term referring to a particular kind of approach to the analysis and design stages of systems development. A formalised version of the use of structured systems analysis principles is known as a 'structured systems methodology'.

Any such methodology must incorporate two components;

1. A set of TOOLS, TECHNIQUES and MODELS for recording and analysing the existing system and the user's new requirements, and for specifying the format of the proposed new system,

2. Some kind of overall FRAMEWORK, indicating which tools are to be used at which stages of the development process, and how they inter-relate.

This of course applies equally to the more traditional approaches as to structured systems analysis. In order to begin our description of the latter, it is perhaps best to posit a somewhat loose definition:

> **A modern approach to the analysis and design stages of the systems development cycle, adopted in order to overcome the weaknesses shown by the more traditional approaches.**

Such a definition clearly needs to be explored in much greater depth, and in order to do that it is necessary to examine the concepts and weaknesses of the more traditional approaches.

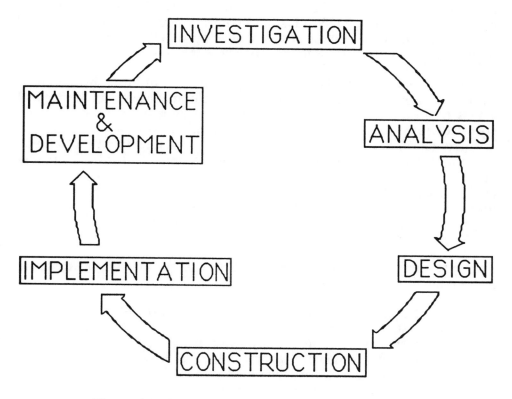

Figure 1.1 Traditional Systems Development Cycle

1.1 The Traditional Approach and its Weaknesses

In one sense the older methods of developing systems followed a much more standard pattern than the modern structured ones (examples of which can be quite diverse). This conformity stems from the creation in the 1960's of a set of standards, supported by techniques, documents, checklists, etc by the National Computing Centre of Great Britain. A level of flexibility was built into these standards in that DP departments were encouraged to adapt them to their own particular needs, perhaps for example using a sub-set of the documentation, and tailoring it for different types of system. The NCC method became incorporated into a six-week Basic Systems Analysis course, designed and marketed by the NCC, and on which students were subject to examination and award of a certificate of competence by the

British Computer Society. These courses were run by technical colleges all over the country, and provided a nationwide standardisation of approach. Within a few years most of the major organisations employing analysts had switched to a version of the standards, and as a result, most of the large operational systems (eg payroll, stock control, accounts, etc.) were created in this fashion. Many of these systems, developed in the early 70's, are still in operation, bearing witness to the efficacy of the approach.

This traditional NCC method, which incidentally is still used by a number of organisations, is based on a systems development cycle of six logical stages (Figure 1.1). These stages are covered by a series of studies and specifications, each of which has a standard format with formal documentation and checklists of tasks. The structure involves detailed examination and recording of the existing system, an analysis process involving a careful examination of the model of the existing system, and a design stage, where a 'full systems specification' is created. For the implementation stage, tasklists of sub-stages and project management guidelines are provided, giving full and thorough coverage of all aspects of the development process.

There are many important and useful tools used in the method, but these are perhaps the most representative (Figure 1.2):

1   **The System Flowchart**, which gives a high-level view of the system, showing the main documents used, how they enter the system, how they pass from department to department, and how the reports and replies leave the system to go to the users. This type of diagram is created for the existing system and also for the new system.

2   **The Procedure Flowchart**, which charts the sequence of the processes occurring in the system. Separate charts are produced for clerical and computer processes, and again this tool is used to analyse and record both the existing and proposed systems. The clerical flowchart format owes much to traditional O & M techniques, and includes the tracing of particular documents through tasks, offices and departments, whereas the computer flowcharting techniques are those used from the early days of programming.

3   **The Computer Run Chart**, which shows how programs are grouped together to form suites, and indicates which programs must be run, in which order, the input and output files used for each, etc. This tool is clearly for use in the design of batch systems. There are however, in all of the more recent versions of the material, several on-line systems counterparts.

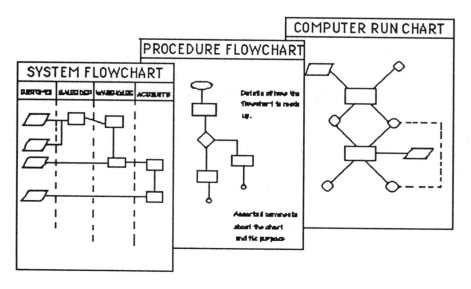

Figure 1.2 Examples of Traditional Systems Tools

It must be pointed out in fairness to the NCC method that it covered a much wider area than any of the current structured systems methodologies attempt to do. The course and material included training in interviewing, in physical file design, in various aspects of project management, and a basic understanding of how a business operates. On the other hand, the structured methodologies tend to concentrate solely on the tools and skills used in modelling the existing and proposed systems, and as a result do not comprise full systems analysis training. This fact has been overlooked by some organisations, who have tried to substitute structured methodology training for the much more thorough NCC course, to their cost!

Having credited the traditional approach for its universality and (given the era in which it was developed) for its control and organisation, it must be admitted that it has not always been found to be effective. In fact there has been increasing evidence of systems delivered late, seriously over budget, and with essential user needs not satisfied.

In recognition of this, the NCC, in the latest version of their material have abandoned many of the traditional tools in favour of structured systems analysis alternatives.

The main weaknesses of the traditional approach have been diagnosed as
lack of control,
lack of user involvement,
the use of inappropriate tools.

1.1.1 Lack of Control

Evidence shows that when using traditional approaches to development, early estimates of time and resource usage can be wildly inaccurate, especially when dealing with very large systems. This is particularly important because estimates are often regarded by the user as promises, and the success or failure of a development will therefore often be judged according to whether the final costs are within those estimates. For example, if a system takes ten months to develop and it was originally estimated to take six, then the development may well be considered as something of a failure (irrespective of whether the system itself gives satisfaction). If however the same system has been estimated to take one year, the project may well be seen as a triumph for the development team (even if there are a number of teething troubles).

There are problems not only in making plans and estimates, but also in taking accurate measurements of work completed in order to check with the agreed schedule. The investigation, construction and implementation stages of a project can be very complex, but at the same time may be relatively easy to measure in terms of person-days; (for example the number of interviews likely to be required can be fairly accurately estimated, and a time can be allowed for each). However, the analysis and design stages (the ones with which we are most concerned when studying structured systems analysis) are much more difficult. The time and resources needed depend on the size and complexity of the problem, not of the system. They are also dependent on the skills, experience and creativity of the people who are to solve the problems, and this can be extremely difficult to estimate for. It is somewhat similar to the task of estimating how long it might take for a carriageful of commuters to each individually complete the Times crossword!

1.1.2 Lack of User Involvement and Understanding

There is of necessity some user involvement in the investigation stage of the project, but often it is a passive one; the users are the only people who know how the existing system works, so the analyst must question them in order to get the information. In the analysis and design stages, users have little or no involvement. All too frequently, at the end of the design stage they are confronted with a massive unreadable systems specification document, full of technical jargon. When they find that they cannot understand it in any detail, they often blame themselves for this, and eventually accept the specification blindly, trusting to the 'computer people' to handle the implementation. From this point on the users are in an increasingly more vulnerable position. Demotivated by slippage and teething troubles, they gradually withdraw their commitment to the new system, and this culminates in their receiving a system which is not at all what is expected, nor what is needed.

1.1.3 Use of Inappropriate Tools

There are three aspects to this. Firstly, the communication tools used in traditional systems analysis provide poor quality documentation. It is a mistake we have often made in the past to provide the same documentation for the user or owner of the system, and for the system builder, who will take the specification and turn it into a database schema and a series of programs. One only has to think of the architect, who provides very detailed scale drawings and instructions for the builder, whereas for the prospective owner an entirely different visual impression of the finished version is provided, perhaps in the form of a scaled-down model or water-colour sketch; the builder's instructions would mean nothing to the owner. The architecture metaphor is particularly apt when discussing structured systems analysis; modern structured methodologies are about creating paper models of the existing and required systems, and it is generally felt that graphical versions of systems are easier for the user to understand.

The second aspect relates to the fact that the analyst and designer are always aiming at a moving target. The company, and the study area within it, are human activity systems, and as such are constantly growing. A major project may take several years to develop, and during that time the objectives and requirements of the organisation may well change. The concept of a 'freeze date' is used in most cases, preventing any changes to the specification after a certain point in the development. However, this can cause problems. Sometimes when a system has been installed and the user identifies extra requirements, the cost and even the feasibility of the necessary amendments may have been adversely affected by the design approach chosen for the original system. What seems to the user to be a minor alteration can cut across the whole structure of the existing design.

The third point relating to the use in traditional approaches of inappropriate tools concerns the changing nature of data processing. When most of these tools were devised, DP systems were almost all batch, and dealt for the most part with operational aspects of the organisation (eg order processing, stock control, etc.). Nowadays the majority of systems are on-line, and a large number of them relate more to management information than operations. There are also different attitudes to the use of hardware and the storage of data. In other words, the industry has moved on, and that fact needs to be reflected in the use of new, more appropriate tools.

So these are the main problems which the structured systems analysis approach has to overcome, and they should be borne in mind during the following more detailed examination.

1.2   The Structured Systems Analysis Approach

There are five main aspects of modern structured methodologies which are worth more detailed discussion, a discussion which should provide us with a fuller and more workable definition of the term 'Structured Systems Analysis'. These are:

1. The formalisation of a set of **tools** and **techniques**, identifying the best type of tool to handle each situation.

2. The standardisation of the **structure** or **framework** of the approach, laying out the essential stages inherent in the analysis and design processes.

3. The distinction between, and the utilisation of, **'physical'** and **'logical'** views of the system.

4. The provision of a substantial role for the **user** in the development process.

5. The use of the inherent **structure of the data** as the basis of the proposed design.

1.2.1 The Tools and Techniques

Structured systems analysis methodologies make use of a number of tools and techniques, which are used to build models of both the existing system and the proposed system. These models help the analyst with the recording, the analysis, and the design, and they provide a full definition of the requirements in a form that is clearly understandable, not only to the designers and implementers, but to the owners and users of the system.

The commonest tools associated with structured systems analysis are
Data Flow Diagram
Entity Model
Relational Model
Hierarchical Function Chart
Entity Life History

Data Dictionary
Structured English
Walkthroughs.

There are others, like for example checklists and forms, used in specific methodologies, and some of the above list are known by names other than those used here. All of these tools will be mentioned in the following pages. A selection of them will be discussed in great detail, and the techniques and skills necessary to make best use of them will be taught.

1.2.2 The Framework of a Methodology

A structured methodology consists not only of a group of modelling tools, but also a method of integrating and interfacing the products of the different models. Each model is a view of the existing or proposed system, and somehow these many different views have to be co-ordinated to give a meaningful requirements specification. A framework consists of a series of steps and stages, backed by forms and checklists, which force a standardised approach to the development process. Some of the enforcement takes the form of a dependency, whereby the product of one step is an essential input requirement for the next.

The advantage of such an approach is that the tools and models support and cross-check each other, providing a much more reliable system. It also means that the steps in the development process can be more easily measured in terms of their products, and therefore controlled. Lack of controllability is one of the main weaknesses of the traditional approach.

There are of course some disadvantages to using a strict framework, resulting in a very formalised approach. It can for instance take much of the flexibility from the analyst, limiting the scope for devising really creative solutions. This however is not always seen as a disadvantage by DP and user management, who are prepared to sacrifice the small possibility of a quick, cheap, brilliantly intuitive solution, for the greater probability of an acceptable solution provided within time and budget (albeit a greater time and larger budget). Data processing projects are high-risk ventures, and this can be seen as a sensible piece of risk management.

Systems development methodologies, whether supplied by Software Houses or built in-house, can be classified according to the level of prescription in their framework. Figure 1.3 shows a continuum based on the rigidity and complexity of systems methodology frameworks, ranging from the TOOLBOX approach, where the use of tools is left very much to the judgement of the analyst, to the COOKBOOK approach, where all decisions on tools and methods are pre-defined in the methodology standards. A number of the more popular structured methodologies are

1 3

listed on the diagram, and their approximate position on the continuum is suggested. In particular, the **Modus** methodology from BIS Applied Systems Ltd. is generally considered to be close to the centre, because of its balance of technical and business ingredients.

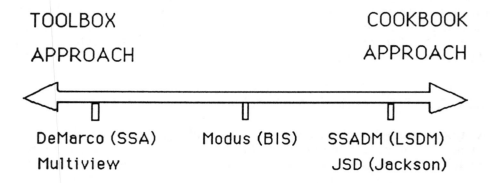

TOOLBOX APPROACH

COOKBOOK APPROACH

DeMarco (SSA)　　Modus (BIS)　　SSADM (LSDM)
Multiview　　　　　　　　　　　JSD (Jackson)

Figure 1.3　The Toolbox - Cookbook Continuum

Whereas a simple Toolbox approach is ideal for the development of small systems and works well in the hands of experienced and creative analysts, for very large systems, where many analysts of differing levels of experience and ability are involved, some form of Cookbook approach is essential.

However, most organisations have projects in their portfolio which vary considerably in size and complexity, and therefore require different development approaches. Such an organisation needs sometimes to use a Toolbox and sometimes a Cookbook. A number of the methodology suppliers are now recognising this need, and are building more flexibility into their frameworks, but there is a danger in this of losing some of the benefits of completeness and control.

1.2.3 Logical and Physical Models

Perhaps the most important advance in systems development practice brought about as a result of structured systems analysis is the formal separation of the 'physical' and 'logical' views of the system.

A physical model of the system is one which describes how the system performs its tasks, who does what, where it is done, how long it takes, etc. It is concerned with the physical constraints placed on the system by the owners, users and designers, in order to make it operate.

On the other hand, a logical model ignores physical constraints; it is only concerned with what functions are required by the system, and what information is needed to carry them out.

In simple terms, the PHYSICAL modelling techniques are used in the *investigation* of the existing system, and in the *design* of the new, whereas the LOGICAL models are used for *analysis* of the system requirements.

The distinction between the two concepts of logical and physical is a major factor in all structured systems methodologies, and the subject will be re-iterated in this book time and time again.

1.2.4   The Role of the User in the Development Process

Structured systems analysis fixes a substantial role for the user in the development of the system. The models used in the early stages have been designed and chosen to be as understandable as possible to the non-technical user, and users are encouraged to cross-check the analyst's findings in structured walkthroughs and presentations. It is the graphical nature of the models that is considered to be the major factor in their user-friendliness.

In many methodologies, users are given the opportunity to be involved in the building of the models as members of the design team, while some methodologies even force the user into an involvement at various points in the multi-stage development process.

1.2.5 System Design based on Data Structure

In several of the most important modern systems methodologies the design of the new system is based heavily on the view of the requirements obtained using Data Analysis modelling techniques. These techniques identify the natural relationships between the different types of entity about which information is held, and check whether the proposed processes are able to access the information they need.

The argument for using the data structure (as opposed to the process structure) as the main framework on which to base the new system is that the data structure is more stable. When the system owner or user wishes to make changes to the system, it is nearly always to enable better use to be made of existing information rather than to obtain more information from extra sources.

To facilitate this, it is suggested that the natural relationships inherent in the data should be incorporated in the design, and that the system's data can then be thought of as comprising a massive 'information warehouse'. Individual processes can then simply be designed to access it in any way required, and any process changes should cause minimum disruption to the rest of the system.

This partly overcomes the 'moving target' problem, mentioned earlier as one of the weaknesses of the traditional approach. More importantly, it makes the process of changing the system during the 'maintenance and development' stage of the system life cycle much easier and cheaper; overall, this stage is almost always more expensive than the original development, so savings made here can be quite substantial.

The use of the data structure as the basis of the system design also makes it easier for the data processing function to grow within the company, through the use of a database strategy.

1.3   Summary

So these are the main points and arguments in favour of a structured systems analysis and design approach. To sum up briefly, one can say that:

Structured systems analysis is a modern approach to the problem of analysing and designing a business computer system.

It involves the use of a group of tools and techniques, and these are integrated by means of a structure or framework of stages and steps.

It involves the building of paper models (in graphical form rather than in text) of both the existing and required systems. These models are used to communicate both with the user and with the proposed builders of the system, the designers and implementers coming after.

The final point in this summary relates to the limitations of structured systems analysis. It is **not**, and does not claim to be, a complete method for system development. It does not include, for example, interviewing techniques, cost-benefit analysis, project management, hardware and software procurement, and a number of other skills and techniques that normally come under the systems analyst's aegis.

Essentially, no matter how 'cookbook' a methodology might be, *it is not a substitute for logical thinking, analytical skill, and creativity.*

2   PROTOTYPING

At this point it is worth introducing, and discussing the merits of, a relatively new type of approach to system development, that of prototyping. It is important because it impacts on the way the analysis and design stages are conducted, and also because it is proving its value within organisations of all sizes. It must be borne in mind that the concepts and techniques involved in prototyping are still developing, so that any definition at this stage is still likely to be a general one. However, it is felt that any modern systems methodology must attempt to take advantage of these techniques, so some kind of working definition is essential.

2.1   Definition

Prototyping can be defined as

> **building a physical working model of the proposed system, and using it to identify weaknesses in our understanding of the real requirements.**

This suggests that prototyping should be used somehow during the SSA stage of 'requirements analysis', perhaps along with the 'paper' models of the data flow diagram and the entity model. It can also be used at a later stage after a relatively detailed investigation, analysis and design, to provide a version of the system which can be adjusted, tuned and optimised.

There are two major different approaches to the practice of prototyping. These are known as **Rapid prototyping** and **Evolutionary prototyping.**

Rapid prototyping involves creating a working model of various parts of the system at a very early stage, after a relatively short investigation. The method used in building it is usually quite informal, the most important factor being the speed with which the model is provided. The model then becomes the starting point from which users can re-examine their expectations and clarify their requirements. When this has been achieved, the prototype model is 'thrown away', and the system is formally developed based on the identified requirements.

On the other hand, Evolutionary prototyping takes place after a more careful investigation, and the methods used in building the prototype are more structured. The reason for this is that the Evolutionary prototype, when built, forms the heart of the new system, and the improvements and further requirements will be built on to it. It is not 'thrown away' as is the case with the Rapid version.

Prototyping has always been a standard technique in the general engineering field, but until recently it has not been feasible in software engineering because the development of code took such a long time. However, with the advent of fourth generation languages, the production of code and the iterative process of amendment can be performed very quickly, allowing an early version of the system to be put together and tried out on the user before the proper system is built.

2.2   The Fourth Generation Environment

In order for prototyping to be conducted properly, it requires not only a fourth generation language (like FOCUS or NOMAD), but also a whole environment of software surrounding it, providing what is often called a 'fourth generation environment' (4GE) or an 'analyst workbench'.

Figure 1.4 gives an illustration of such an environment, with a data dictionary system at its centre and a database management system attached. The screen formatter and the report generator tools have been available for some time, but gain extra power when incorporated into a full workbench. The fourth generation language itself usually comprises two languages (or two modes of the same language): firstly, a very high level, mostly non-procedural language for the end users, to enable them to make enquiries and perform minor updates, and secondly a much more complex and rigorous language, to be used by DP specialists (ie. the analyst/programmer or prototyper).

The other tools marked on the figure by dotted lines are optional and of varying importance. One of these, the graphical design tool, allows for the capture of data flow diagrams and entity models on screen, and the automatic cross-checking of these models by means of an active data dictionary. Such tools, examples of which are 'IEW' from Knowledgeware, and 'Automate Plus' from LBMS, are growing in importance all the time. They are known as CASE tools (CASE being an acronym for 'computer assisted software engineering').

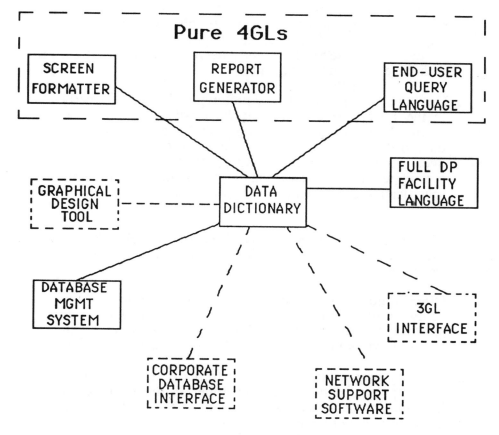

Figure 1.4 The Components of a 4GL Environment

2.3   Methods of Prototyping

Different prototyping strategies are adopted by different organisations, and within an organisation, depending on the type of application being developed.

1. One strategy is to develop a prototype of only the most important parts of the system, ignoring for the moment the details on the periphery (eg. the exceptions and controls). Very commonly a single-user prototype is developed for an eventual multi-user system.

2. Sometimes only the screen interfaces are prototyped. This allows users to get a feel for the eventual system, and to help to tune the screen layouts to their satisfaction.

3. Often prototyping is done only on one or more sub-systems of the whole application. Commonly this occurs where the application involves a mixture of on-line and batch sub-systems, because on-line lends itself to the prototyping technique much more so than batch.

4. There are two situations where prototyping is used very early in the system development cycle. Firstly it can be used during the feasibility study stage to help convince users of the worth of such a system. Secondly, it can also be used even when the intention is to purchase a package; the prototype can serve as a major part of the requirements specification and the request for proposals.

The tactics and techniques used in prototyping are still being developed, but there are several points which can be made to illustrate how the task is performed currently.

1. Often prototyping is assigned as a specialist role, and the analyst/programmers carrying out the task are known as 'prototypers' or 'system builders'. They work in small teams of two or three, to minimise communication problems.

2. In many organisations, the only formalised tool used by prototypers is data modelling, either in the form of an intuitive approach (as described in chapter 6), or by using Third Normal Form analysis (chapter 7).

3. Prototypers, particularly those who adopt a Rapid Prototyping approach, make use of pieces of old prototype systems and standard general purpose modules as well as the fourth generation statements. This 'component engineering' method allows for a prototype to be built very quickly using a 'cutting and pasting' technique. However, the product is not usually in a form which can satisfactorily be used for evolutionary purposes.

4. Some organisations have set up small prototyping centres, either as part of their Information Centre, or within the main Data Processing department. This enables the specialist staff to maintain 'pattern books' of screens and systems, so that potential users can browse through them, looking for appropriate styles of screen layout, or models on which they would like their new system to be built.

2.4   Types of Project Suitable for Prototyping

Prototyping is not a suitable technique for all types of application project, and research is currently being carried out to identify the circumstances where it proves most and least successful. However, some general guidelines as to which types of system are most suitable can be given at this stage.

It has been found that prototyping is very effective in the analysis and design of on-line systems, especially for transaction processing, where the use of screen dialogues is much in evidence. The more the interaction between the computer and the user, the more benefit can be obtained from building a quick system and letting the user play with it. On the other hand there is usually very little advantage in prototyping a batch system.

Systems with a great deal of calculation in them, again are not really suitable for prototyping; the coding necessary to do the work may take more time than is normally set aside for prototyping. Also, the main advantage from prototyping, ie. the user being able to see the product early enough to make requirement revisions, does not apply in this situation.

There is some dispute as to how useful the prototyping technique is for very large systems. In most organisations the high level of standards required for the development of a large system have not been specified for the use of prototyping. On the other hand, because of the speed at which a prototyped system can be developed, many systems which might formerly have been classified as very large might now be classed as only medium sized.

The last aspect of a system which makes it suitable for prototyping is the atmosphere and environment in which the project is to be done. The users must understand and be in favour of the approach, the software 'workbench' facilities must be available, and the hardware and software environment in which the live system is to run must be compatible with the environment in which the prototype is developed.

2.5   The Conflict

There are however problems in trying to make use of both structured systems analysis techniques and prototyping in the same project. These problems stem mainly from the difference in the philosophy of the two approaches. While it is true that both can be seen as mechanisms for improving the process of systems development, both approaches have evolved from different assumptions of how systems development should be done, assumptions which may be said to contain elements of incompatibility.

Structured systems analysis and design methodologies use modern effective modelling tools, but base their frameworks on the traditional systems development cycle, which is split into the main stages of Investigation, Analysis, Design, Construction, and Implementation. Each stage is broken into easily measurable units, and a stage is normally considered to have to be completed before the next stage starts. This provides a clear demarcation between different processes, and allows for the use of specialist staff at different stages: (eg. business analysts during the investigation and analysis stages, more technical systems designers during the design, and programmers during the construction). The reasoning behind the use of these formal stages is the same as for a major project in any other field (such as architecture or engineering): ie. that each component, once constructed, will underpin the components to follow. The later that an error or misunderstanding is discovered, the harder and more expensive it is to put it right.

This is considered particularly true of the programming task, which can be the most expensive and time-consuming of all. In any large development project, it is essential that a full, accurate and explicit program specification is provided by the analyst before programming begins (the programmer only being authorised to carry out the limited task of translating requirements into a computer language).

On the other hand, the use of a fourth generation environment and prototyping techniques reduces the criticality of this programming stage, in that much fewer instructions take less time to create, and therefore can be changed more readily. These higher-level languages can also be used by less specialist staff, for example the business analyst, and even by the user, cutting down on the need for a formal program specification as communication.

The concept of prototyping cuts across the development cycle stages, as each prototype includes elements of all four stages, Investigation, Analysis, Design, and Construction. The earlier definition of prototyping would suggest that it should take place during the 'requirements analysis' stage of the project. However, at that point in most structured systems methodologies, no physical design has been considered, so building a physical model should not be possible! Many established methodologies are now having prototyping techniques retro-fitted into their structure, without any real changes being made to the existing stages and tasks. Early evidence suggests that this approach meets with limited success.

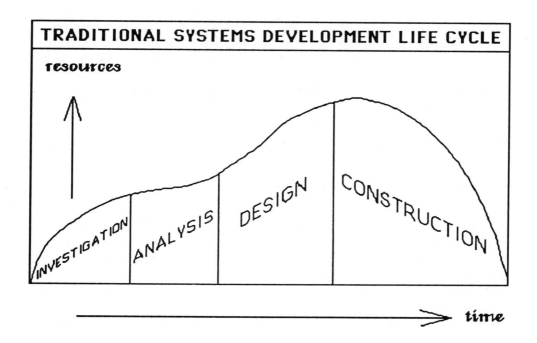

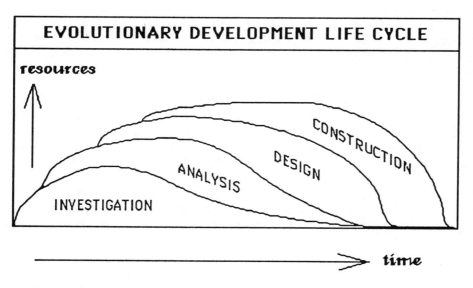

Figure 1.5 Traditional Systems Development versus Evolutionary
Systems Development: Comparison of Life Cycles

The other important element in the prototyping technique is iteration; the process of building part of the system, checking it with users, then altering or even rebuilding, and rechecking, etc. until it is considered satisfactory. This again clashes with the structured systems analysis approach, where each step is fitted into a dependency schedule, and consists of a sequence of tasks, each of which can be checked and signed off. Here we notice the contrast of design strategy at the heart of the conflict. Whereas SSA methodologies (in general) put forward a one-off constructive view of development, prototyping supports the evolutionary approach, in which modifications are considered to be the norm. This evolutionary approach in design and construction is then naturally continued in the succeeding implementation and maintenance stages.

The use of prototyping for systems development has led in places to a change in the user's view of what is required: a user is now able to go for a more advanced system with greater functionality and greater human-computer interaction. Fourth generation environments lend themselves particularly to the development of on-line systems, specialising, as they do, in dialogue design facilities. Moreover, many fourth generation language routines, after they have been developed, become standard procedures or components in future systems, requiring only slight tailoring in order to fit the new system requirements. This 'component engineering' approach means that the user can often be given, at a very early stage, a number of options in prototype form, and be asked to make a decision accordingly. Being able to see a version of the proposed system gives users a much clearer view of their requirements. They are much more able to see the potential of the proposed system in prototype form than in the form of the 'paper' models provided by SSA.

On the other hand, SSA is much better at analysing the 'logical' requirements of the system, providing a much deeper understanding of the business, opening up a larger range of design and implementation options, and giving the user wider choice. There are undoubtedly systems for which a prototyping approach would be unsuitable or unduly restrictive.

The problem remains, how do we take advantage of the mature, thorough, risk-minimising approach of SSA, and at the same time reap the benefits of a more quickly produced, and more user-oriented system, as provided using a prototyping approach?

2 EVOLUTIONARY SYSTEMS DEVELOPMENT

A simple definition of an Evolutionary Systems Development Methodology (or EDM) might be as follows:

> An approach whereby the initial design proposals are put forward in the form of a physical working model. The analysts, working in partnership with the users, gradually improve and develop this prototype of the system until it meets a level of acceptability decided on by the user. When this happens, **the prototype becomes the new system**.

This chapter examines some of the most important ideas in the theory of evolutionary systems development, illustrates these with examples from literature and practice, and provides a list of essential concepts which must be incorporated in a good EDM.

An important secondary purpose of this chapter is to establish the 'pedigree' of EDM; ie. to illustrate that the approach is accepted and recommended by many of the leading thinkers in the industry, and has proved itself in practice in many different environments.

One of the most authoritative voices in IT is that of James Martin, who, in his book *An Information Systems Manifesto,* criticised much of today's research into methods of improving systems development, for trying to solve the wrong problem.

> "The important problem is how to migrate from conventional programming and the old development life cycle to development methodologies which are fast, flexible, interactive, and provide provably correct code; methodologies in which interactive prototyping replaces formal, voluminous specifications which must be frozen; methodologies which are automated; methodologies with which end-users, managers, specifiers, implementers and maintainers can interact without mis-matches"

Martin's seminal book suggests ideas and directions, gives pointers and expounds philosophies. He leaves it to others to fill in the details and provide the 'nuts and bolts'.

1 RAPID PROTOTYPING METHODOLOGIES

Many organisations who make use of Rapid prototyping (as opposed to Evolutionary prototyping) have managed to adjust traditional or structured methodologies to incorporate the new tool. These approaches are mostly based on a model put forward by Bernard Boar in his important book, *Applications Prototyping*. Boar saw the purpose of the prototype as to provide a rigorous and user-checked requirements specification, and he proposed that the iterative process used in the prototyping technique could be treated as a small cycle within the full development cycle, occurring during the analysis and early design stages. Figure 2.1 gives an example of this kind of structure. The model has been improved and elaborated by authors such as Dearnley and Mayhew, and Law (see Bibliography), who has even produced a version which can incorporate evolutionary development.

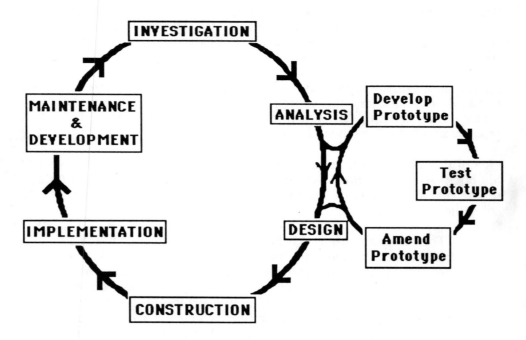

Figure 2.1 A Rapid Prototyping Development Cycle

There are many practical examples of the adoption of such a methodology, and it is perhaps worth examining one in some detail. Softwright Systems Ltd., a software house based in Chertsey, Surrey, make use of an in-house methodology which supports rapid prototyping. Figure 2.2 illustrates the methodology, and shows a comparison with a typical traditional development approach.

Traditional	SSL Approach
1. Feasibility Study	1. Briefing (few days discussion etc. with users and DP dept.)
2. Systems Investigation & Analysis	2. Prototyping (3 to 4 weeks max.)
3. Design & Specification	3. Review (walkthroughs of prototypes, revisions etc. for 2 weeks or so)
4. Code (can use 4GL for this stage)	4. Development (1 to 9 months)
5. Testing	↑
6. Implementation	Benefits from early completion
7. Documentation	↓

Figure 2.2   Softwright Systems Ltd. Prototyping Methodology

The four stages of the SSL development approach may seem to be somewhat insubstantial, but they do incorporate all the essential components of a development project.

1. During the 'briefing', the SSL team (usually 3 or 4 members) talk to the client's managers and users, and form an early picture of the main requirements. They also talk to the client's IT department, to ensure that any design proposal will be compatible with existing IT policy and strategy. At this point it is possible to ascertain whether the project is suited to the SSL approach, and if it is not (a rare occurrence) then an alternative approach is considered.

2. The prototyping process involves creating the most important parts of the required system. This is done using data analysis modelling techniques, setting up a data dictionary, formatting screens with a painter, then writing the procedures in a 4GL. The language normally used is 'Sourcewriter', a very powerful Cobol pre-processor.

3. The review stage comprises many iterations of run-through of the prototype, during which adjustments are continually being made. All levels of user are very heavily involved in

this process of fine-tuning the requirements and constraints which make up the final product of the stage, the functional specification.

4. For the development stage, the prototype code is 'thrown away', and the new system is freshly coded using the lessons learnt during the prototyping phase. This stage of the project is similar to that used in the traditional approach, except that some of the sub-processes can be completed more quickly

Experienced analysts within the company are of the opinion that using the SSL method can save as much as two thirds of the development time of a more traditional development. It does have to be borne in mind that the company chooses the projects to which they apply the method, and that it uses a small number of analysts who are all experienced in the particular hardware/software environment!

2 TRADITIONAL SYSTEMS DEVELOPMENT LIFE CYCLE

The methodologies described above all adhere to the traditional life cycle for the development of a system. However, as already suggested in chapter 1, this cycle can in some circumstances prove to be an encumbrance: Daniel McCracken, another world-renowned expert, has even suggested that it can be positively harmful! But can we manage without it?

Many senior systems developers take for granted the fact that the SDLC should always be used, but as Kit Grindley points out in his book *4GLs A Survey of Best Practice*, the stages of the cycle are there for historic reasons, and were developed to solve problems which may no longer exist. The case has been put by many authors, some at great length. This is a precis of the argument:

1. When computers started to be used in business systems, hardware storage capacities were severely limited. Information processing had to be carried out within the straitjacket of magnetic tape, sequential file access, sorts and batch processing. This made the development of systems, and the systems themselves, unduly complex.

2. In particular, given the nature of early generation languages, the coding of programs was a highly skilled task which required the use of specialist coders. Specifications had to be provided for these coders, and these had to be as accurate as

possible because of the great expense involved in re-coding after error. This meant that the specification had to be complete before it was given to the programmer, and this brought into existence two successive development stages, **Design** (or Specification) and **Construction**.

3. Because the design had to be completely correct before the coding began, the design could not be carried out until all the requirements and constraints had been identified. This necessitated a detailed **Investigation**, and the findings had to be subjected to careful **Analysis**. Both of these tasks required different skills from those of the more technically-oriented designer or coder.

4. Only when all programs had been completed could the new system be introduced. This meant that there had to be a separate **Implementation** stage following construction.

5. Because the whole development could take a great deal of time, the user requirements might change after the investigation but before the delivery of the system. Such changes could not be incorporated in the planned delivery, because this would cause delay; in the worst case continual changes could mean that the system would never be delivered! The solution to this problem was to set a 'freeze-date' some time at about the end of the analysis stage, after which all changes would have to be saved for a later released version of the system. Of course, this meant that there was still important development work to be done even after the system had gone live, and this was the start of the **Maintenance** stage.

In the time since the traditional SDLC was invented (or, more accurately, evolved) the whole IT industry and environment has changed. The argument put forward by those favouring an alternative life cycle is that the reasons for the original separation of the stages have now disappeared.

1. The capacity of disk and main memory has increased dramatically with the years, and database techniques have greatly simplified file access. So the complexity caused by early hardware constraints is no longer there.

2. Advances in software, in the form of 4GLs, have again reduced the complexity of systems development, to the point where a separate dedicated 'coder' is no longer necessary. 4GL code can be produced by the analyst, sometimes even by the user.

3. If the analyst does the coding, there is no longer the need for a detailed formal specification to provide communication between two separate specialists. Design and construction can overlap.

4. If, through use of the prototyping technique, the user can be involved in the design and construction stages, then there is no need for the analyst to attempt to capture a full and complete knowledge of the user's business in the formal requirements specification. The analyst and user can work in partnership, each with their own area of expertise. If it is necessary during the design to check details of some of the requirements, the user can be on hand to provide the appropriate information.

The most important factor in this major change which has taken place is that producing code is no longer the inhibitively expensive and time-critical activity that it used to be. This means it is no longer absolutely essential that the first attempt is correct, because now we can afford to 'throw away' incorrect code and re-write it correctly.

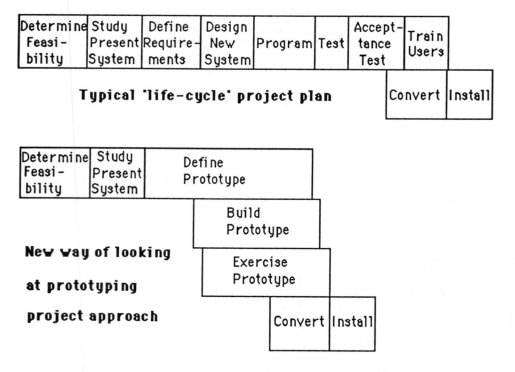

Figure 2.3 Comparison of Life Cycles (Lantz 1984)

However, even if the restrictions that the SDLC places upon development are the result of history (Grindley calls them the 'mistakes of history') many important, even essential, concepts have been built on top of the SDLC. These include many structured systems and project management techniques, and if we are now going to lay the SDLC aside, we must be sure that these concepts are incorporated in whatever we replace it with!

3 EVOLUTIONARY DEVELOPMENT METHODOLOGIES

Once it is accepted that the traditional SDLC is no longer essential to the systems development process, then true evolutionary approaches can be considered. One of the earliest full descriptions of an evolutionary systems development based on the prototyping technique is given by Kenneth Lantz in his book *The Prototyping Methodology*. Again the difference between evolutionary and pre-specification methods is highlighted, and Figure 2.3 (taken from Lantz) illustrates the advantageous overlap of stages.

In the more detailed description of the methodology, Lantz makes surprisingly little use of the more popular structured techniques, preferring instead the more traditional schematic diagrams, supported by extensive walkthroughs. The DFD is mentioned but its use is played down, and only very limited coverage is given to the whole field of data analysis.

On the other hand, he places great stress on the project management aspects of evolutionary development, and on the very important concept of 'exercising' the prototype. There is a natural tendency when considering the prototype verification process to assume that the user simply 'looks at' the prototype and gives an informal opinion on its quality and correctness. In fact, if a prototyping approach is to be used to best advantage, the whole process of examining a prototype and presenting conclusions must be formalised and standardised. From the project management point of view, it must be possible to make estimates of the number of prototypes to be built, the anticipated number of versions of each prototype, and the expected time involved for each iteration.

Lantz also provides one of the earliest lists of benefits to be obtained from using an EDM. Other authors have extended this as further benefits have come to light, but the original version still presents a powerful argument. The benefits are:

>   Involving, committing and satisfying users
>   Decreasing communications problems

31

Decreasing development costs
Reducing operational costs
Reducing calendar time for development
Producing the right system first time
Cutting manpower requirements during development.

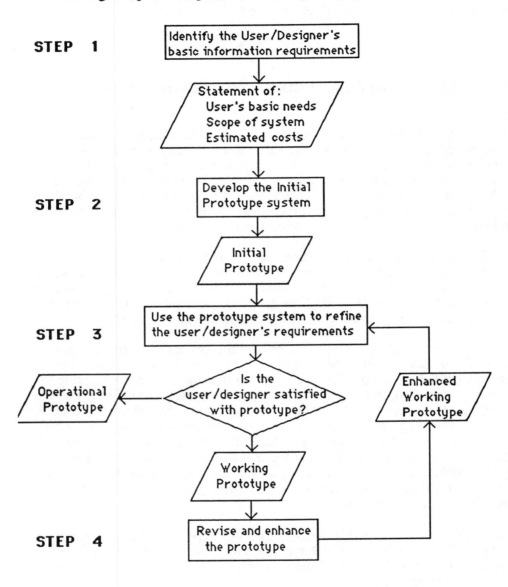

STEP 1 → Identify the User/Designer's basic information requirements

Statement of:
User's basic needs
Scope of system
Estimated costs

STEP 2 → Develop the Initial Prototype system

Initial Prototype

STEP 3 → Use the prototype system to refine the user/designer's requirements

Is the user/designer satisfied with prototype?

Operational Prototype

Enhanced Working Prototype

Working Prototype

STEP 4 → Revise and enhance the prototype

Figure 2.4 Milton Jenkins' Evolutionary Development Methodology
(Jenkins 1985)

Perhaps the best known and most publicised example of an evolutionary methodology is the one put forward by Milton Jenkins, in his article *Prototyping: A Methodology for the Design and Development of Application Systems* as early as 1985. In his approach, Jenkins puts forward the idea that a single prototype will be developed for the system, and iterated until it is of a satisfactory standard to be considered 'operational'. His work has been particularly important in highlighting and formalising the roles and relationships of the user and analyst within the prototyping activity. Surprisingly perhaps, he made use of traditional program flowcharting techniques to illustrate the iterative aspect of the methodology (Figure 2.4). However, the diagram certainly emphasises the clarity and simplicity of the methodology framework, which has since become the model on which other EDM structures have been based. More will be said of Jenkins' valuable contribution to the subject later in the book.

Again there are many practical examples of evolutionary systems development methodologies being used within business organisations. One such example is an in-house approach used by the glass manufacturers Pilkington Brothers of St Helens, Lancashire. The methodology is based on what the company describes as the 'iterative/collaborative' system development cycle, and this is illustrated in Figure 2.5.

As can be seen from the diagram, the Pilkington approach is quite a formal methodology, having clearly delineated stages and sub-stages. In fact the early and late stages of the cycle appear to be similar to the more traditional SDLC; however the analysis, design and construction processes are done simultaneously through a series of iterated prototypes.

Unlike in the Lantz and Jenkins models, the final prototype does not automatically become the live version of the system. Decisions are made during the later stages of prototyping whether to consider a package, re-write some of the system in a 3GL, or to carry the prototype into the implementation stage; (the latter being by far the most common selection). This emphasises the flexibility of the approach, which can encompass within an evolutionary development philosophy the option of a traditional pre-specification development.

There is very little use here of structured analysis and design techniques, other than a relatively high-level hierarchical function diagram. Most of the specification and documentation is in text form. The particular language used is FOCUS from Information Builders, a powerful and well-established 4GL, selected by Pilkingtons after productivity trials (the results of which were very impressive).

The complete commitment of the users is essential in the Pilkington approach, as all of the stages involve joint user/analyst activities, and there are a number of disadvantages in this for some projects. Nevertheless, the method has been used successfully on a number of very large systems, and staff who have used it are convinced that it is the most effective approach for systems development.

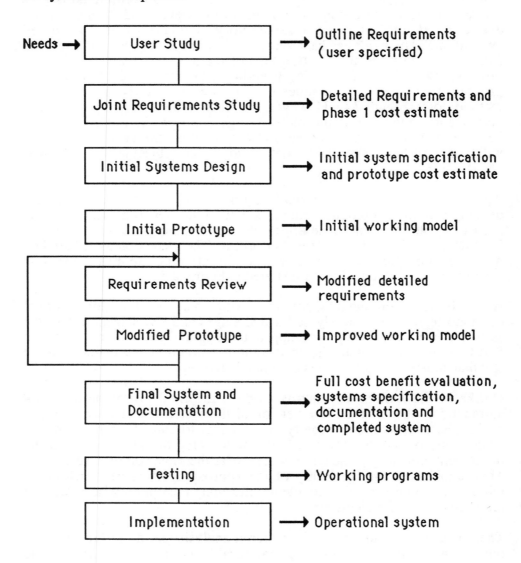

Figure 2.5 The Pilkington Iterative/Collaborative Cycle

The Pilkington case is only one of a number of possible examples of in-house evolutionary systems development methology. There are also important examples of such methodologies devised by software houses and marketed to clients, one of the most popular being 'ESM' from Southcourt Ltd. of Cowfold, West Sussex. This approach provides facilities for a complete top-down analysis of an organisation's information needs, from strategy down to operations, and shares many similarities with the 'Information Engineering' structured systems methodology (including the extensive use of modern data analysis techniques).

4   INCREMENTAL DELIVERY

Another important aspect of modern evolutionary development theory is the early implementation, where possible, of prototyped parts of the system, even though there may still be further work to be done before all requirements are satisfied. It is not unusual within a prototyping environment for the user to put an initial prototype to practical use while waiting for a more developed version. Clearly there are circumstances where this would be impossible, and even when it is possible there are likely to be serious risks. However, the intelligent and committed user who has worked on the prototype will appreciate these risks better than anyone else, having identified its weaknesses and suggested changes. The user may decide that a 'flawed' system is better than no system at all, and besides, aren't all systems flawed to some degree?

Putting aside the wisdom (or otherwise) of implementing an unproved prototype, there is still a case for examining whether the proposed development can be carried out so that some parts of the system can be delivered early, while work on the rest of the system continues. This is what is traditionally known as a 'phased' development.

Tom Gilb, in his recent book, *Principles of Software Engineering Management* , points out that the phased approach is not used nearly as much as it could be. He suggests that the whole project should be planned on the basis of being able to provide a series of 'incremental deliveries'. This means that parts of the new system may be in place early, and earning benefits for the company, benefits which will help to pay for the development of the later parts!

Gilb also observes that the use of a prototyping approach makes splitting up the project into small deliverable units a much simpler and more natural task, and in fact many of the ideas from the book are used in modern evolutionary development methodologies.

The work of George Rzevski and his colleagues at Kingston Polytechnic also emphasises the importance of incremental delivery. They have developed an evolutionary design methodology, known as 'EDM', which partitions the future system into a series of relatively self-contained parts, and develops and implements them as pilot systems. The term 'pilot' is used in preference to the term 'prototype', though, in practice, there is little difference between the Rzevski definition of a pilot, and what is described in this book as a *deliverable* version of an evolutionary prototype.

5   THE ACADEMIC CONTRIBUTION

Most of the progress in the field of modern systems analysis and design has been industry-led rather than academically inspired. However, some important research relating to evolutionary development and prototyping has been carried out. An overview of this research is given in *Software Prototyping in the Eighties*, an Open University IT briefing, and a more detailed survey is contained in Pamela Mayhew's thesis *An Investigation of Information Systems Prototyping*.

One of these important areas of research has been into what are referred to as 'classifications' of prototype, the classifications being based on the different purposes for which a prototype might be used. An extended list of five categories of prototype has been put forward by Law, based on earlier work by Floyd (see Bibliography), and it is as follows:

> **Exploratory**, to assist in the clarification of requirements
>
> **Experimental**, to find the solution to a particular problem
>
> **Performance**, to check whether a solution can handle the workload
>
> **Organisational**, to test a solution within a proposed environment
>
> **Evolutionary**, to be developed into an operational system.

A practical conclusion from this work is that, during the development of a system, a prototype may be used at different stages for different purposes. This means that the same prototype at different stages of its evolution will be used by the analyst to test and prove different things, and it introduces the concept of a prototype **version** as distinct from an **iteration** (an iteration simply being a change to the prototype instigated by the user).

One other strand of research which has produced valuable results, specifically within the field of software engineering, is that conducted by M. M. Lehman and L.A. Belady. In their book, *Program Evolution*, they summarise all the main arguments and evidence concerning the evolutionary development of software built up since the early 70s.

One particular observation from this work which is worth highlighting concerns the fact that not only does the **design** of the program evolve, but the **program** itself continues this evolution throughout the rest of its life. The description of the working life of the program as the 'maintenance' stage is a misnomer; the program does not deteriorate in the traditional engineering sense, the user's requirements change and the program must 'evolve' to bring it in line with the revised specification. As there is really no difference between the kind of iterative adjustments made during the evolutionary development and the 'amendments' made during the working life of the program, the hitherto universally accepted dichotomy between 'Development' programming and 'Maintenance' programming is no longer relevant in any evolutionary approach! (More is said on the subject of system maintenance in chapter 13.)

There are three major areas of research where prototyping techniques are used, and which are related in some way to the development of business systems, but do not yet have firm practical results in the field. These areas are:

HCI (Human Computer Interface)

Formal Languages (remember Martin's quote about 'provable' code)

Knowledge-based Systems (where evolutionary methods dominate).

No doubt in time, benefits from this research will feed into the mainstream of systems development, and therefore any framework that is outlined now must be flexible enough to incorporate new ideas when this time comes.

The most effective research in the short term is in the form of collaborative projects between business and educational establishments. These have the advantage of providing a live environment in which the researchers can test their theories, and at the same time enable the business organisation to reap the immediate benefits of modern advances. There are many examples of such work, one of the most notable being the COMET project for the Royal Dockyards, involving research staff from the University of East Anglia .

Another instance of this co-operation between academic and business interests is the 'Systemscraft' methodology, which is used in the rest of this book as a full example of an evolutionary systems development approach. Systemscraft was originally developed at the City University, London, and was tested in a number of organisations, in particular at a company called BSL International of Bromley, Kent, who currently market a version of the methodology known as 'EASE' (Evolutionary Approach to Systems Engineering).

One last point is worth making concerning the academic contribution to EDM development, and that relates to user involvement in analysis and design. During the late 70s and early 80s, a great deal of research was done by Enid Mumford at the Manchester Business School into the subject of 'Participative Design'. This explored methods of involving the users at all stages in the development of new business computer systems, and emphasised the need to design jobs which provided staff with the kind of rewards and satisfactions that they required. This research was considered highly in academic circles, though a number of the techniques proved to be expensive in staff time, and the changing economic climate and new 'spirit of the age' of the 80s made the approach somewhat hard to justify. However, the advent of 4GLs and prototyping techniques have meant that this extensive involvement of the user is not only financially feasible but is actually beneficial! As a result, many of the research findings from the participative design research can be, and are, applied in modern EDMs. (One example of this is how to deal with a prototype which is to be operated by a large number of users.)

6   ESSENTIAL CRITERIA FOR AN EDM

From the description in this chapter of the main theory and concepts of evolutionary systems development, it is possible to construct a list of the characteristics one would expect to find in a good EDM. These are as follows:

1. **Overlap of Analysis, Design and Construction Stages**; the prototype must be designed and constructed, but its purpose is analysis.

2. **Limited Modelling of the Existing System**; a much greater emphasis on the early identification of the 'logical' requirements, so that quick prototypes can be produced.

3. **Partnership and User Responsibility**; the analyst and user each bringing their own expertise to the development task, and sharing the work and responsibility.

4. **Formalisation of Prototype Boundaries**; as a system development can involve many small prototypes, it is important to delineate each prototype to avoid ambiguity and duplication.

5. **Evolving Levels of Functionality**; as each prototype evolves during the development, there should be a gradual and controlled increase in its complexity and purpose.

6. **Early Implementation**; every opportunity should be sought to deliver parts of the system as early as possible.

7. **Flexibility and Scalability**; it must be possible to use the methodology for projects of varying size and nature, and to 'tailor' the methodology for each different project.

7   CONCLUSION

Evolutionary development methods have themselves evolved over the last ten years or so. In one sense the evolutionary approach was tried once before in the history of IT: in the early days before there were proper standards for systems analysis, programmers would try to 'evolve' their programs by simply coding until everything came right. It has taken a long time to live down the disasters that this caused, but now at last we have the well-thought-out techniques which will enable us to evolve systems properly.

There will still of course be reservations; the problems of project management and the dangers of partial delivery of systems will send a shudder down the spine of any conscientious DP manager! However, it is in the nature of business systems to evolve, and they have been doing so for many thousands of years before computers were invented. A business system is in essence a human activity system, and as such is organic. There will always be some computer systems which must be constructed like cathedrals, but there are others which need to be allowed to grow like flowers. A good evolutionary development methodology must be able to deal with both, and all conditions in between!

3 THE METHODOLOGY FRAMEWORK

The purpose of this chapter is to introduce a 'nuts and bolts' example of an evolutionary development methodology. The actual tools and techniques used in the methodology itself will be discussed in great detail in parts 2 and 3 of the book, but before exploring these, it is important to consider some of the broader aspects of the approach, such as:

> the overall structure of the methodology,
> what its main components are,
> why it has been chosen as the example,
> how it compares with other EDM examples,
> how it relates to the full systems development process,
> whether it satisfies the 'essential criteria' from chapter 2.

It was mentioned in the previous chapter that there are now a number of working EDMs in the market place and within major organisations, carrying the message of evolutionary prototyping, and proving their worth in many different environments. However, such methodologies are still very new, and there has not yet been time for an industry standard to evolve, or even for a 'market leader' to emerge. Because of this, the example methodology has been chosen for its comprehensive coverage of the key elements of evolutionary systems development theory, rather than for its market share.

The name of the methodology is 'Systemscraft' (though versions which are customised for specific organisation are usually given different names). It was devised originally as a method of teaching the use of structured systems techniques, and was 're-engineered' to incorporate EDM principles, initially to provide designs for the best use of the FOCUS 4GL. Since then it has been used with NOMAD and with Oracle 4GEs, and versions have been produced that are 'tailored' to make use of Yourdon and SSADM structured methodology standards.

It should be remembered that the methodology is being put forward here not just as an example, but as a kind of prototype, which users (DP departments) can adjust to their particular requirements. Any readers who are exploring this book with a view to making use of an EDM as part of their company's systems development strategy should bear in mind that this is just one physical manifestation of the logical principles of evolutionary systems development. There are many options and variations which could be used to construct a version of the methodology to fit more closely with their existing standards and requirements.

1 THE SYSTEMSCRAFT METHODOLOGY

The Systemscraft methodology is an attempt to bring together the critical elements of structured systems analysis and prototyping, providing in one systems development method the advantages of both approaches. A minimal tool set has been used, and every attempt has been made to preserve the method's simplicity and the flexibility, placing it towards the 'toolbox' end of the previously mentioned continuum. Emphasis has also been placed on its ease of understanding for the user, particularly in the early stages.

1.1 Its Place in the Full Development Cycle

The methodology basically consists of a number of integrated modelling tools and techniques for the analysis of the business and the design of the computer system. However, the use of a prototyping approach, particularly one of the evolutionary type, automatically infers the inclusion of the construction and implementation activities. (This is one of the important differences between an EDM and a structured systems methodology: the product of the structured methodology is a 'specification' for later construction, whereas the product of an EDM is a fully implemented system.)

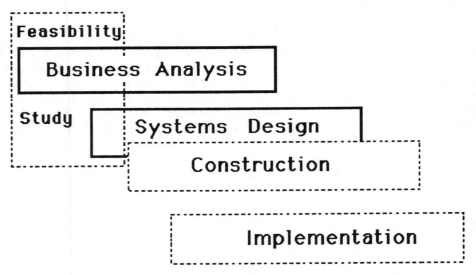

Figure 3.1 The Systemscraft Methodology
Stages in the Evolutionary Development Cycle

Figure 3.1 illustrates the different stages involved in an evolutionary development project, and gives an impression of the level of overlap possible between these stages. Only the two major stages, that of the

Business Analysis and the Systems Design, are covered in depth; the Feasibility Study stage is examined in the final part of the book, and aspects of Construction and Implementation are addressed as they arise in the text. 'Investigation' is not shown as a separate formal stage, although obviously it must take place in order for there to be material for analysis. In this methodology (as in several of the most advanced structured methodologies) it is not considered necessary to build models of the physical existing system, though there may occasionally be an advantage in using a model to clarify a complex physical procedure. The view is taken that most of the information about the existing physical system does not need to be 'captured' in the traditional way during the investigation, because much of it is irrelevant to the new system, and because the users, who are experts in that system, are part of the development team, and are constantly on hand to provide the relevant parts when required!

The most important point to take from the diagram is the fact that the two major stages will be worked on concurrently for a large part of the project; it is not necessary for all parts of the business analysis to be complete before the designing starts. This means that prototyping, which is seen as part of the design process, can begin at a relatively early stage (perhaps within the first few days).

It is common practice for a methodology to be described using its own tools and models. Such a description exists for this methodology, but for the sake of brevity and simplicity, it is best described in this chapter with the help of a diagram known as a 'roadmap', which illustrates the modelling tools used during each stage; (Figures 3.2 and 3.3).

2   THE BUSINESS ANALYSIS STAGE

The purpose of this stage is to produce a logical analysis of the existing system, extracting from it the implicit requirements of the business, which must obviously be taken into account in the design of the new system. Figure 3.2 illustrates the products of this logical analysis, collectively known as **The Business Model**. There are four extremely well-established structured modelling techniques used in this stage, and they are used in their simplest form. They are:

> the Hierarchical Function Diagram
> the Data Flow Diagram
> the Entity-Relationship Model
> the Relational Model

It is also the purpose of the stage to include in the business requirements any extra facilities which the user identifies that are not already in the current system. These are incorporated in a second group of integrated diagrams, the Required Business Model which in practice is often a slightly adjusted copy of the earlier model. It is comparatively rare that the user requires extensive new processes or data; usually the main concern is to get the current requirements handled faster and more accurately. Of course, there are some developments for which there is no existing system! Obviously in this case the Business Model is built from the user's expected requirements.

2.1   Analysing the Business Functions

One of the most difficult problems facing the junior analyst is that of identifying the logical requirements behind the physical reality of the current system as it exists. In order to make this task easier, the methodology forces a hierarchical decomposition of the business functions within the study area (the Business Function Diagram). This emphasis on the function rather than process makes it difficult for the analyst to lapse into recording how processes are physically carried out. This idea is carried forward into the next modelling activity, as the functions of the BFD become the processes of the Data Flow Diagram. The standards used for the DFD are those of DeMarco, but with a few minor additions; every attempt has been made to keep these early tools as user-friendly as possible. The two models, the BFD and the DFD, cross-check each other, and are iterated until a logically simple and aesthetically satisfying model of the functional requirements is arrived at.

One other important aspect of the use of the Business Function Diagram as the first modelling tool is that the analyst can often identify whole integral functions, which can then be investigated, analysed and designed separately from the other functions. This can mean that the analyst is sometimes able to progress one 'leg' of the function hierarchy through to the design stage while analysis work is still being done on other parts of the system. This is a key point in the strategy of providing early prototypes.

2.2   Analysing the Business Information Requirements

In order to examine in detail the information requirements of the system, two further models are used; they are referred to here, again for simplicity reasons, as the Data Model and the Relational Model. The Data Model is basically an entity-attribute-relationship model, arrived at through a top-down approach; (the terms *data model* and *entity model* are treated as synonymous throughout this book). The Relational Model is built from the attributes identified in the earlier model, and put through the 'normalisation' process. These two models are obviously used to cross-check each other, but they are also integrated with the functional models in that the entity types on the final version of the data model become the 'data stores' on the final version of the data flow diagram.

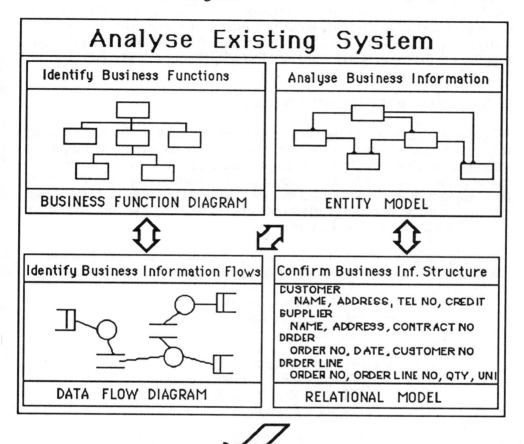

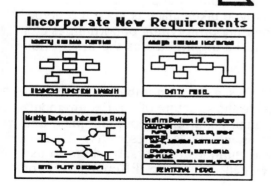

Figure 3.2

Systemscraft Methodology

Business Analysis Stage

It is accepted that these four models alone cannot constitute the full documentation for a requirements specification; they must be supported by some explanatory text, data dictionary entries, descriptions of DFD low-level processes, background information, etc. However, the level and content of this supporting material is a matter of judgement between the analyst, the user, and the project manager. All that the methodology supplies are the essentials to guarantee that a full analysis is conducted.

3   THE SYSTEMS DESIGN STAGE

Whereas the Business Analysis stage deals purely with the logical view of the system, and is never concerned with the way requirements might be satisfied, the Systems Design stage (Figure 3.3) involves the immediate examination of options for implementing these business requirements using the computer. It is recommended that the design stage for some parts of the system should be started as early as possible in order to take full advantage of opportunities to build prototypes.

The design process uses, as its main input, all or part of the Specification of Requirements constructed during the analysis process, and consisting of

> Business Function Diagram
> Business Data Flow Diagram
> Entity Model
> Relational Model
> Supporting Documentation, including;
>> Process Specifications
>> Physical Requirements Chart
>> Data Dictionary.

As well as this, there needs to be a full and continuous dialogue conducted with user and owner throughout the design process. This may take the form of further fact-finding interviews, walkthroughs and presentations. The most important way in which this dialogue takes place is via a prototype, and the methodology is geared to providing prototyping opportunities at a number of stages of design. These will be highlighted as the description progresses.

Whereas the analysis process comprises a small number of tools and techniques, all of which were essential to the structure of the Requirements Specification, the design process, though apparently more complex, and containing a wider range of techniques, is actually less prescriptive. There is much greater flexibility as to which techniques to use and where to use them. Also, most of these techniques can be used to different levels of complexity:

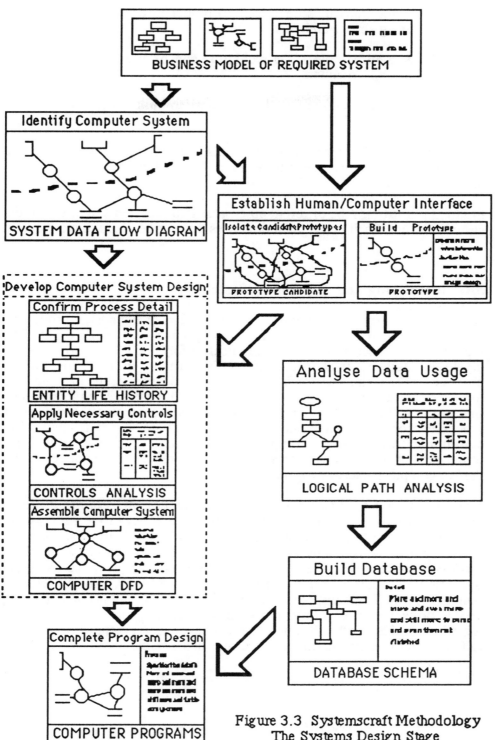

Figure 3.3 Systemscraft Methodology
The Systems Design Stage

47

1. Sometimes it is necessary to model to a great depth, using strict standards,

2. Sometimes the same technique can be used in a less rigorous manner,

3. Sometimes it may not be necessary to build that kind of model at all.

All the skills necessary for using these tools and techniques should be in the modern systems analyst's repertoire, even though they may not all be used in every project.

Figure 3.3 illustrates the different processes, steps and tools involved in the systems design. Each of the main processes is described here in more detail.

3.1   Identifying the Computer System

As can be seen from the diagram, this is the first of the design processes, and it uses the Business Data Flow Diagram as its main input. The modelling tool used in the process is itself a form of DFD, known as the 'Systems' Data Flow Diagram, and it is built by splitting the logical processes of the Business DFD into physical processes, some of which are to be carried out by the computer and some by the users. The thick broken line passing through the centre of the diagram separates the computer processes from the manual ones. However, there are sure to be data flows communicating between computer and manual processes, and these represent the documents, forms, reports or screens which make up the human-computer interface of the system.

Individual designers may work on different function areas of the Business DFD, each area perhaps representing only a small part of the full system. The designers will use experience and judgement to identify one or more ways in which the computer could be used to help in the carrying out of the processes on the DFD. These different options will be modelled using the Systems DFD tool, then discussed with the owner/user, and a preferred option agreed. These options may include a batch or an on-line approach, and various levels of computer involvement, ranging from a simple recording of the fact that business actions have been taken, through to the full computer implementation of the actions themselves.

3.2 Confirming the Inputs and Outputs

Once the user has agreed to a computer option for part of the system, the designer is able to identify areas of the system which might suit a prototyping approach. It is normal practice to build a number of small prototypes of different parts of the system, rather than one large prototype of the whole. Good potential prototypes occur where the human-computer interface is complex, and involves on-line dialogue: (batch processes do not normally make worthwhile prototyping opportunities).

The separate prototype candidates are marked on the Systems DFD, and can be issued to individual analyst/programmers, who will build their prototypes using details from the requirements specification, copies of documents from the old system, and long discussions with the user. The Systems DFD will indicate which logical data stores are made use of by the processes being prototyped, so 'dummy' files can be constructed to allow simulation of these accesses. It may be possible to construct a first prototype very quickly, and then to deploy the user's reactions in honing it to the required level of quality.

Almost inevitably, the prototyper must have access to a fourth generation environment in order to function successfully.

3.3 Analysing the Usage of the Database

This process is concerned with the way in which the data accesses made in the systems processes (as highlighted in the individual prototypes) match with the logical data model built during the analysis stage. The data-accessing processes are mapped onto the data model, showing the detailed pattern of data usage, and providing source material for the physical design of the file/database system.

Although the 'roadmap' in Figure 3.3 illustrates only one form of model for this task, there are actually three different types of model used in the analysis of data usage. The first of these, the Path Analysis Diagram, plots the data entities which need to be accessed by each process in the proposed computer system. The second, known as the Navigation Model, imposes all these individual paths onto the original data model, giving a high-level picture of the system's total data usage. The third type of model is called a Data Usage Chart, and it shows the details of all accesses for each entity table in the data model (an entity table being a potential data file). The combined effect of the three models is to turn data access information about each process into data access information about each logical file.

3.4 Developing the Computer System Design

This is the most complicated of the design processes shown on the diagram, and in fact is made up of three separate sub-processes. These sub-processes together provide the detailed definition of the system's computer programs. They can be applied following the development of initial prototypes, and they fill out the detail of the user's requirements through the use of more complex modelling techniques and through the provision of more advanced prototypes. In cases where a prototyping approach is not seen to be useful, these sub-processes produce a full specification of user's physical requirements, and provide clear models from which the sub-systems, programs and subroutines can be constructed.

It is an important characteristic of the methodology that it should be flexible and scalable enough to support all forms of business system development. Obviously some of the smaller developments will not require the same level of design effort as the larger ones, so many of the techniques from these sub-processes are to be used only when the nature of the development requires it.

3.4.1 Confirming the Computer Process Details

The purpose of this sub-process is to make sure that none of the less obvious aspects of the proposed new system have been overlooked. The DFD is a very useful tool for analysis and design, but it lacks a full self-checking mechanism. To enable us to check for the completeness of the Systems DFD, we use a tool known as the Entity Life History.

The approach involves examining each of the entity types from the Data Model, and creating a sequenced list of events where actions will affect it. For example, each entity must have an event which sets in motion the creation of an entity. Equally, some event must trigger the action to delete an individual entity. Each attribute in the entity may have its value inserted or modified by one or more actions. All of the event-oriented actions affecting each particular entity type are then checked against the processes on the DFD; does every action belong to a DFD process? If not, then the DFD needs to be reconsidered.

In some systems, the order in which events relating to a particular entity occur can be of critical importance. For example, before a payment can be made, there must not only have been an invoice, but also a confirmation of delivery must have been received; if payment is made without this confirmation, an error may have occurred. Where it is considered necessary, a 'status indicator' attribute is added to an entity type, to hold details of the last action to have taken place. Error procedures for wrong sequences must also be designed, and incorporated into the other models making up the specification.

3.4.2 Applying Necessary Controls

This sub-process involves examining the whole system shown in the Systems DFD (including the non-computer parts) and identifying which controls need to be applied. This activity should involve the commitment and participation of the internal auditors as well as the prospective users of the completed system, because it is concerned with identifying areas in the system which are vulnerable to

> Theft,
> Loss of company assets,
> Mis-informed business decisions,
> Excessive costs,
> Compromise of proprietary information.

When such an area is found, decisions must be made as to the level of controls needed to counteract the weakness, and this will depend on how critical the situation might be for the organisation, how likely it is to occur, and how much the control mechanisms may cost.

The method used consists of examining the exposed parts of the system, as shown on the Systems DFD. These potentially vulnerable outputs will be

1. the data that flows to outside agents,
2. the data stores inside the system; (there may be unofficial access to these).

The possible situations and circumstances in which damage could occur to each of these outputs is examined, by tracing back through the DFD model. As each point of weakness is identified on the model, some control activity must be decided upon, and will be included in the specification.

3.4.3 Grouping the Computer System Components

So far, all the processes of the system which are to be handled on the computer have been treated as individual activities. They are shown on the Systems DFD as being related to other processes involved in carrying out a higher-level business function. At this stage however, it is necessary to combine the individual computer processes into computer sub-systems, program suites, program runs, programs, modules and subroutines, and these combinations are unlikely to be along purely functional lines.

The main tool used in the grouping of the computer system components is known as the Computer Data Flow Diagram. This is a fully physical DFD, in contrast to the logical Business DFD and the semi-physical Systems DFD. The technique is equally capable of modelling on-line and batch situations, and acts as a replacement for both of the more traditional tools, the Computer Run Chart and the Network Diagram. This multiple use of the DFD within the methodology fits in with the general policy of minimising the number of different types of modelling technique in the analyst's 'toolbox'.

When the individual processes are pulled together as modules in a larger program, special extra modules may have to be created in order to provide selection dialogue and access control. Again, these modules may be put to the user in the form of prototypes, to ensure acceptance of the interaction format.

3.5 Designing the Physical Database

This describes the process whereby the file/database designer creates the data definitions for the proposed live system, and sets up the master file structures ready for implementation. The following information is needed for this to be carried out:

1. The user's performance constraints,
    in terms of h/w and s/w, response times, etc.

2. Details from the analysis of data usage carried out
    earlier; ie.
    the Data Model
    the Relational Model
    the Path Analysis Diagrams
    the Navigation Model
    the Data Usage Charts.

The designer will normally start by applying what are referred to as 'first-cut rules' to the logical data model, to convert it into a preliminary set of files in accordance with the particular file-handling software being used by the organisation. From then on, these files will be 'massaged' and optimised until they fit with the performance requirements of the system.

To begin with, this optimisation process is likely to be a paper exercise, whereby the data usage models are adjusted and re-calculated. However, in time-critical systems it is often necessary to build a 'performance prototype' of one or more of the most important files, to make sure that the required file constraints can be met.

Obviously the exact procedures adopted by the file/database designer are heavily dependent on the particular software to be used, and on the hardware that is available. As a result it is not possible to be specific about this process; all that can be provided are guidelines and examples.

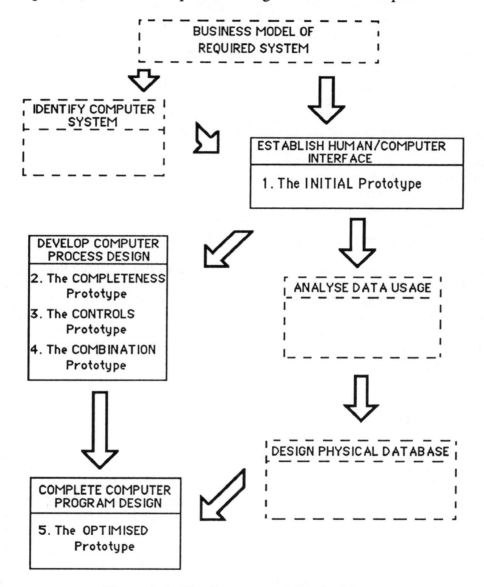

Figure 3.4   The Systemscraft Methodology
Evolving Versions of Prototype

3.6 Completing the Program Design

The detail of this process is dependent on the software to be used, the type of system involved, and on the analysis and design approach taken in the earlier stages. For example, it may involve 'tidying up' an evolutionary prototype to take account of database optimisation, the re-writing of a prototype in a third generation language for performance reasons, the creation of computer programs from non-prototyped specifications, etc.

Normally, this stage will be the culmination of the evolutionary development process. The final versions of the systems prototypes will be provided, and these will be iterated until they are in a form that is completely acceptable to the user. At this point they will become the delivered system!

Figure 3.4 illustrates the gradual evolution of the system through five different levels of prototype, each of these versions having a different purpose, and containing more functionality than the previous one. This provision of a series of versions, each one subject to several iterations, means that the user is not only in constant touch with the development of the system but is contributing regularly to its construction.

In spite of the strong emphasis on prototyping in this methodology, it is recognised that in many business computer systems there are large and complex calculation processes, for which the use of 4GLs and prototyping are completely unsuitable. In these circumstances, the programming would have to be carried out using rigorous structured programming techniques, and by a specialist programmer. Such programming techniques are obviously outside the scope of this methodology.

4 SUMMARY

The Systemscraft methodology has been designed to provide a full set of modelling tools for the analysis and design of business computer systems. In the design stages in particular, there are a variety of tools, not all of which will be utilised in every system. For example, when the system is small, when it is logically simple, when the response time is not critical, or when the system's information is not security-sensitive, some of the design stages may require only token treatment. Figure 3.5 shows a roadmap of the methodology in its minimal form; this includes only those tools and stages which are essential for any systems development.

Even when a modelling technique is to be used, there may be different levels of complexity and detail to which it can be applied. Whereas under-utilisation of systems development tools can result in poor quality systems, their over-utilisation leading to waste of resources can be almost as bad.

The analyst in charge of the development must use judgement in deciding exactly what techniques are to be used, and to what level they are to be applied, in order to balance the two risks. This level of flexibility and scalability is claimed to be one of the main benefits of the methodology.

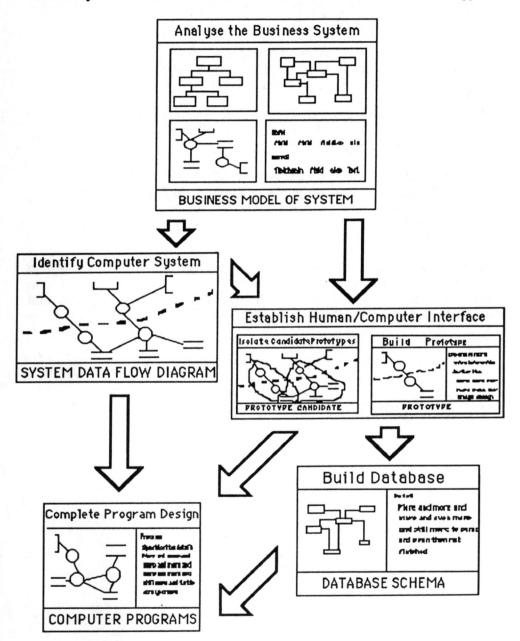

Figure 3.5 The Systemscraft Methodology in its Simplest Form

5 APPROACH TO THE REST OF THE BOOK

In this first part of the book, the principles and theory of evolutionary systems development have been discussed, and an example methodology has been outlined. At this point, the reader should decide upon a strategy for tackling the rest of the book.

> Part 2 deals with the **Business Analysis** stage, and discusses in depth the various modelling techniques employed during that stage. (Chapters 4 to 8.)

> Part 3 does the same for the **Systems Design** stage, concentrating on the detailed modelling of the computer system. (Chapters 9 to 12.)

> Part 4 returns to the broader aspects of ESD, and discusses criteria for tailoring the approach, and the project management implications. It concludes by summarising the whole subject, and looking briefly to the future. (Chapters 13 to 15.)

Readers who are mainly interested in a managerial view of the topic may wish to continue from the start of Part 4, whereas those who wish to explore or revise structured systems analysis and design skills and techniques may prefer to continue through Parts 2 and 3.

Part 2

4 IDENTIFYING THE BUSINESS FUNCTIONS

THE BUSINESS FUNCTION DIAGRAM

This is the first of five chapters which examine in detail the Business Analysis stage of the Systemscraft methodology. Four of these chapters describe and explain the use of major structured systems techniques, while the remaining chapter places them in context by summarising the stage, and pulling together the important ideas and implications.

Throughout the book, the use of the various structured modelling techniques are illustrated using a simple case study, an order processing and stock control system for the Gentry Shoe Company. This particular case study has been chosen because it represents one of the simplest and best known business situations, and should therefore require no specialist knowledge on the part of the reader. A brief description of the case is given in Appendix 1, but for the most part the diagrams will be self-explanatory.

The first step in the analysis of a system is to identify the **business functions** to be carried out by the system to be developed. A business function is a logical concept rather than a physical one; it describes what needs to be done for the business to be conducted, not where, or how, or by whom it is done. This 'functional' view is only one of a number of views taken of the system during the analysis stage, but it is a particularly useful one at the start of the process. It represents the way that the owner (as opposed to the user) sees the system, and is the most user-friendly of all the models, involving one of the simplest of the modelling techniques used in any methodology.

1   DEFINITION

The Business Function Diagram is a simple hierarchical decomposition of the functions of the system within the study area. Each function is boxed, and if necessary broken down into sub-functions, the number of levels depending on the size and complexity of the system. A statement of **purpose** is recorded on the diagram, (normally a single sentence elaborating the top level function), along with a statement indicating the part of the organisation with **responsibility** for carrying out the functions.

The purposes of the diagram (see Figure 4.1) are;

1. To help identify the scope of the system to be analysed.

2. To help enforce a 'logical' approach to the analysis of the system. The functions identified here are used in a number of later models as potential processes, and the more purely functional these are (ie. the less they are constrained by physical aspects like who is to do it, when and where), the more flexibility will be available at the design stage.

3. To indicate the position of the study area system within the overall system of the organisation. This can clarify responsibilities, help prevent duplication of effort, and identify duplicate and redundant processes in the existing systems.

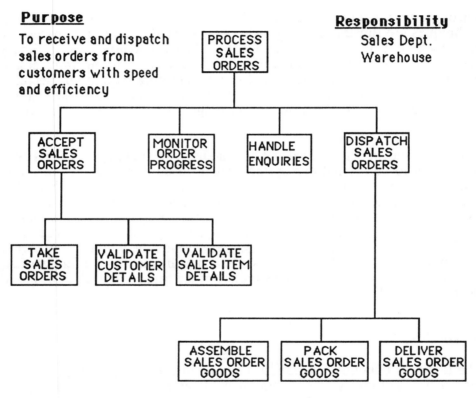

Figure 4.1 A Business Function Diagram

It is important to realise that the analysis process, of which this is a part, is taking place at the same time as the detailed investigation process. Traditional systems analysis textbooks have, quite rightly, stressed the separation of the analysis from the investigation, but that separation does not necessarily mean that the investigation must be complete before analysis starts. The use of the 'top-down' approach in this early modelling technique means that after a brief high-level investigation, the main functions can be identified, and work can go ahead simultaneously on the different function areas. The investigation process is the collection and discussion of the facts concerning the system, and the analysis is the building of the logical models.

There are two forms that the business function modelling process can take, and which of the two is to be used depends on the company's DP strategy, and also on the importance and complexity of the system. The two forms are:

1. The **Standard** form, where the system is considered to exist only within the study area.This approach is taken with small and naturally integral systems.

2. The **Corporate** form, where it is important to position the study within the overall DP approach of the company.

The standard form will be discussed first, and the differences relating to the corporate approach will be detailed towards the end of the chapter.

2 THE FORMAT OF THE DIAGRAM

2.1 Hierarchical Decomposition

The essence of this approach is that any function in its box consists of all functions from boxes connected to it lower down the tree; no more, no less. So, in Figure 4.2, function C consists of functions D and E, (function E consisting of F, G and H). Obviously, the whole system, Function A, consists of B and C, which include all functions further down the tree.

2.2 Levels

There may be any number of levels down through which the diagram may go. However, it is unusual for even a large system to merit more than six levels, and it is quite common for a small or medium sized system to require no more than three.

As a rule, there should never be more than seven sub-functions of one function (ie. children of one parent) on a diagram, as it can make the model harder to assimilate. It can also cause difficulties of over-crowding in some of the later models which are built using the BFD as a base. The problem of too many sub-functions can be solved by the insertion of an intermediate level.

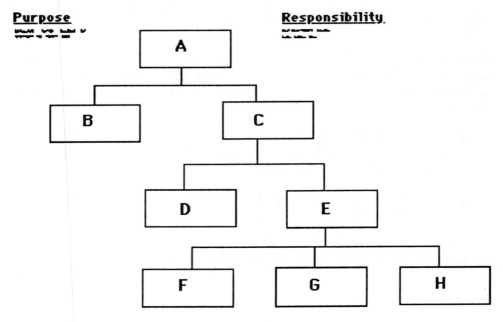

Figure 4.2 The Nature of Hierarchical Decomposition

The diagram should be relatively 'balanced', in that the levels at which the lowest sub-functions are defined should be roughly the same. For example, Figure 4.2 is a poorly balanced diagram whereas Figure 4.3 is well balanced. Another aspect of the need for balance is that sub-functions of the same parent function should be of approximately the same size, complexity and importance.

2.3   Function Names

Each function must be given a unique name, always in the form of a VERB - OBJECT, and the verb should be in the imperative mode.

eg. 'Take Orders', 'Maintain Stock', 'Purchase Goods'

The quality of the names given to the functions is critical to the success of the system. The name should represent as closely as possible the full range of sub-functions for which the function stands. As has already been stressed, this diagram is used as the starting point for a number of later modelling techniques, and poorly named functions can cause confusion, and even error, deep into the development process.

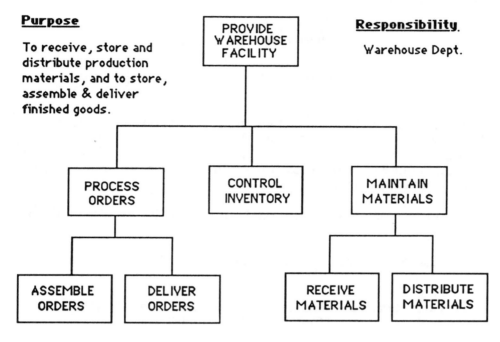

Purpose

To receive, store and distribute production materials, and to store, assemble & deliver finished goods.

Responsibility

Warehouse Dept.

Figure 4.3   A Balanced Diagram

Although, the systems analyst is involved in designing the 'information' system within the study area, many such systems have a real world of activity, of which the information system is simply a 'paper' model for purposes of control. For example, the real world of a warehouse system is to do with unloading supplier deliveries from lorries, moving goods about the stores, assembling customer orders, etc. The paper documents, like the order forms, delivery notes, and picking lists, represent the control system working above this reality, and it is the re-design of this control system with which the analyst is mainly concerned. However, the BFD is a model of the whole business system, not just the control aspect, so function names should reflect the functions of the real world, not just of the information system.

eg. 'Store Goods',   not just 'Record Goods Storage' (although the latter may be a sub-function of the former).

'Assemble Orders', 'Deliver Goods', 'Inspect Materials',

If this approach is not taken, the analyst may design an apparently excellent information system, but which does not work because it does not mirror the reality of the business.

2.4   Identification of Functions

In most circumstances, the functions and sub-functions in a system can be identified intuitively, based on the information obtained during investigation. However, sometimes in complex situations this can prove difficult. In those circumstances, it is worth taking a more formal approach to the identification process.

At the top level of a business a major function will do one of the following three things:

Provide a Product   (eg. 'Assemble Product', 'Manufacture Goods'),

Provide a Service   (eg. 'Repair Customer Goods', 'Purchase Supplies', 'Sell Goods'),

Manage Resources   (eg. 'Manage Accounts', 'Maintain Stock', 'Maintain Personnel').

The analyst should be able to identify which of these applies to the system being studied, and the appropriate name for the function can be discussed and agreed with the owner/user.

In attempting to split high-level functions into sub-functions, a technique known as Life Cycle Analysis can sometimes be of use. This technique involves taking the high-level function and identifying exactly what it is providing for the company. It may, as we have already seen, be providing a product, service or resource. In each case, the analyst can look at four 'life-stages' within the company, and each stage may suggest one or more potential sub-functions.

The life-stages are

1. The identification of a need, or the planning for the provision.

2. The obtention, and/or the implementation.

3. The maintenance and support.

4. The relinquishment or the disposal.

Taking for example the high-level function 'Sell Goods',

The **identification** of need may suggest a function 'Research Customer Requirement' or 'Forecast Sales'.

The **obtention** may suggest 'Take Orders', and the **implementation** may suggest 'Assemble Orders', 'Deliver Orders', etc.

The **maintenance** may suggest 'Maintain Customer Details' and 'Handle Enquiries'.

The **disposal** may suggest nothing immediately, other than an indication that there is a quite low-level sub-function to weed out obsolete customer and order details from company records.

These newly identified functions should give the analyst plenty of food for thought, prompting decisions on further levels of breakdown.

There is a large aesthetic component in the building of Business Function Diagrams. The final product has to be a balance of clarity, simplicity, and accuracy. There are always many alternative interpretations of the business requirements, and the purpose of this modelling technique is to find the most appropriate and effective one, which will also gain the acceptance of the user.

2.5   Lowest-Level Functions

As the analyst proceeds to decompose functions of the model through several levels, a decision must be made as to which functions require no further decomposition. The most important point to make about this may seem obvious; the analyst should stop when there ceases to be any advantage in further decomposition. This clearly will depend on the nature and size of the project.

Experience, including an understanding of the use of the functions in other modelling techniques, should enable the analyst to make this judgement.

4. SUMMARY

The Business Function Diagram is the first modelling tool used in the analysis process. It helps to define the boundaries of the system being studied, and provides components for later modelling techniques. It can be a relatively subjective view of the system, so it is important that a good quality model is produced, and that full agreement is reached with the owner/user.

When some of the later models are produced, the detail of the Business Function Diagram may be 'reviewed', and a revised and improved version may be created.

5 IDENTIFYING THE BUSINESS INFORMATION FLOWS

THE DATA FLOW DIAGRAM

The previous chapter described how a purely 'functional' view can be taken of the system under study. The next stage in the analysis process is to examine in some detail the information needed to carry out the functions identified, and the information to be provided on their completion. The modelling tool used for this purpose is probably the most widely known and commonly used of all structured systems analysis tools, the Data Flow Diagram. In this chapter, the format and general principles of the DFD are discussed, with emphasis on its place within the Systemscraft methodology, and on its inter-relationship with other models.

1 DEFINITION

The Data Flow Diagram is a tool which can be used to support four of the analyst's main activities;

ANALYSIS	The DFD is used to help define user requirements.
DESIGN	It is also used to plan and illustrate options for the analyst and user to consider when designing the new system.
COMMUNICATION	One of the main strengths of the DFD is its simplicity and understandability to analysts and users alike.
DOCUMENTATION	The use of graphical models such as the DFD in the formal Requirements Specification and Systems Design Specification is a major simplifying factor in the production and acceptance of those documents.

The Data Flow Diagram provides a model of the system which gives a balanced view of both the data and the processes. It shows how information flows from one process or function in the system to another. Most importantly, it shows what information must be present before a function or process can be carried out. This emphasis on identification of data requirements, categorises the DFD as part of the analysis rather than the investigation process, and distinguishes it clearly from the more traditional 'flowchart', which indicates the sequence of procedures and flow of control of the process.

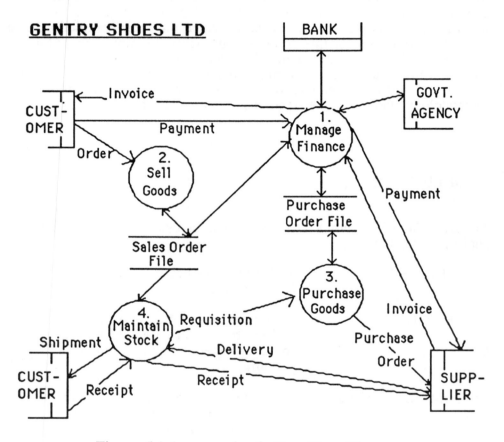

Figure 5.1 An example of a Data Flow Diagram

It is not claimed that the data flow diagram provides a full analysis of the system. It does not, for example, indicate the time element (eg. how long it takes information to get from one process to another). It does not even specify the order in which functions must be carried out (though the order is often evident from the dependency of one function on the product of another). Nor does it indicate the quantities of data involved, the volumes, trends, peaks and troughs; information which is an essential component in the analysis process. So the data flow diagram has clear limitations; these missing views and aspects of analysis must be provided using other techniques.

Figure 5.1 illustrates a business system which has been modelled using the DFD tool. The main functions of the business are identified (in the circles), and the information necessary for each to be carried out is indicated, along with the source of that information. It should be noted that some of the data used by a particular function comes from within the system (eg. from another function, or from a file), and some comes from outside (eg. from customers and suppliers). The same is true for information produced by a process; it may go elsewhere in the system, or it may go to some external organisation. This aspect of the model is important, in that it enables the analyst and user to specify the boundaries of the system.

2 DFD NOTATION

Each of the types of symbol used in the model are described below

2.1 The Process (or Function)

In the diagram a circle is used to indicate a Function or Process. The difference between a function and a process is important, and will be discussed in more detail later, but for the moment it is as well to treat them as synonymous. The use of the circle symbol is a convention, inherited from the early process-based methodologies. Many modern methodologies have adopted other symbols for this purpose, such as the rectangle and the round-cornered square, the argument for change being that it is easier to type text into a box shape. This methodology retains the circle symbol because it helps slightly in the provision of a simpler, clearer and more user-friendly diagram.

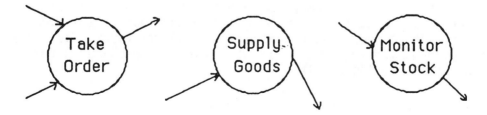

Figure 5.2 A DFD Function or Process

A function described on a DFD must TRANSFORM information. That is, it must change the information from its input in some way, either by reorganising it, adding to it, or creating from it. If in a DFD process no new information is produced, then it is a false process in DFD terms, and the activities involved should be merged with those of a real data-transformation process.

The names given to processes should be unique, and should always be in the form of a VERB-OBJECT, the verb being in the imperative mode; for example, 'Accept Supplies', 'Load Vehicles', 'Store Materials', etc.

In fact, the names for the processes should be the same names as those given to functions on the Business Function Diagram, (chapter 4). The important relationship between these two models will be discussed in great detail later in the chapter.

2.2 The Data Flow

A Data Flow is the passage of information of some sort into or out from a process. It is indicated on the diagram by a line with an arrow on at least one end. The arrow indicates the direction of the information flow.

Figure 5.3 Examples of Data Flows

Each data flow should have a name attached to it. This name may not necessarily be unique, in that the same information may flow into a number of processes, but care must be exercised to ensure that flows of different information are not given the same name. Information which suffers some change of status during a process should be given a revised name on exit to indicate this; (see Figure 5.4).

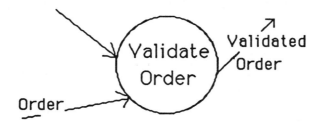

Figure 5.4 A Change of Information Status

Usually (but not always), business information is transported by means of clerical or computer documents, but it is the information which is important, not the piece of paper. The data flows, and the names given to them, should indicate the 'logical' information involved, not the physical documents. The purpose of analysing the existing system is to discover the inherent business requirements, and the use of a specific type of document in the system is unlikely to be essential. The required task could be done in any number of ways, and the objective of our analysis is to open up that potential, to enable the designer to offer alternatives. Our purpose in building the data flow diagram is to help us see beyond what actually happens in the present system, to clarify what functions and information are necessary to fulfil the task.

Some examples of good logical names, and poor physical ones, are given below.

'Order' is a good flow name, whereas 'order form' is bad.

'Payment' is good, whereas 'cheque' is bad.

'Delivery confirmation' is good, whereas 'delivery note' is weaker.

2.3   The Data Store

Data Stores on a DFD represent information which is required to be held over a period of time, in order to be accessible by one or more processes or agents. In physical terms, they might be files of documents held in cabinets, or computer files held on disk. However, it is not the physical medium that is of interest, but the logical information contained therein.

The symbol used for a data store is a pair of parallel lines, enclosing the name of the information stored. The same data store may be included on a page of the DFD any number of times, to help make the diagram as readable as possible.

Figure 5.5 Examples of Data Stores

When a data store is being accessed or updated, there will be data flows indicating the fact. It should however again be stressed how important it is that the passage of 'information' is recorded, and not the physical movement of documents.

It has already been suggested that data flows must have names attached. The only exception to this rule is the flow to and from a data store. If the flow represents an access to, or update of, the standard unit of the file (eg. in the case of a customer file, an individual customer record), then that can be implied by simply leaving the flow blank. However, this is not a hard and fast rule, and the analyst can over-ride it in the interests of clarity.

2.4 The External Agent

An External Agent is a person, group, or organisation outside the study area of the system, but with whom the system has some form of contact. The presence of these agents on the diagram locates the boundary of the system, and pinpoints its relationships with the outside world. It is important to realise that 'outside the study area' does not necessarily mean outside the organisation; if a study of the order processing system is being conducted, then the Accounts department, the Purchasing department, and parts of the warehouse may all be external agents.

Figure 5.6 Examples of External Agents

External agents are a vital part of any system. They are the sources from which information comes to our system, and the recipients of its products. The symbols used are shown in Figure 5.6. (Please note the double line which distinguishes them from internal agents.)

2.5   The Internal Agent

Whereas an external agent name will always be a noun, representing a section, department or organisation, the Internal Agent name will always be in the verb-object form. An internal agent is a function or process within the system being studied, but one which is featured on a different page of the model. Any DFD model is likely to consist of a number of pages, and the information passing between processes on different pages is indicated by this symbol.

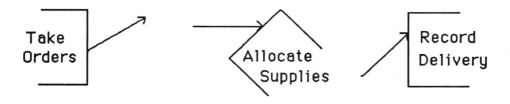

Figure 5.7 Examples of Internal Agents

Both internal and external agents can be duplicated any number of times on the same diagram page, in order to simplify or improve its appearance.

3   LEVELLING

A full data flow diagram of a system under study will almost certainly be too complex to fit on one page of diagram, so the technique of hierarchical decomposition is used to split the diagram down into a number of levels. The top-level diagram (known as the level 0 diagram) consists of the main processes within the system. The content of each of these processes may be expanded onto a full page, where its sub-processes are identified, and their data inter-flows modelled. Each sub-process may in turn be expanded into a full page of its own, and this decomposition can continue down through as many levels as is necessary, (see Figure 5.8).

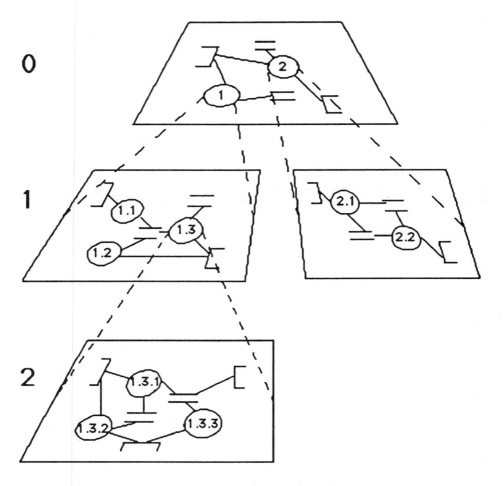

Figure 5.8 Levels of Data Flow Diagram

Each page of the diagram should be given a name. In the case of the level 0 diagram, the name will be that of the whole system, and each lower level page will be given the name of the process of which it is an expansion. It is also sensible to use a numbering system, similar to the one shown in Figure 5.8. In this approach, each process in the top-level diagram is given a number, which is carried down and used, with a subscript, as the numeric identifier of the each of its sub-processes. Both these aspects are illustrated in the example of a multi-level DFD model in Figures 5.9 and 5.10.

The diagrams at the different levels must be checked against each other for consistency. For example, if a process is expanded into a series of sub-processes on a new page, a check must be made that the flows into and out from the process are all present in the lower-level diagram, and vice versa. While performing this check, it should be borne in mind that sometimes the flow at the higher level may be generalised, and may be represented at the lower level by more than one flow. An example of this can be seen in Figure 5.10.

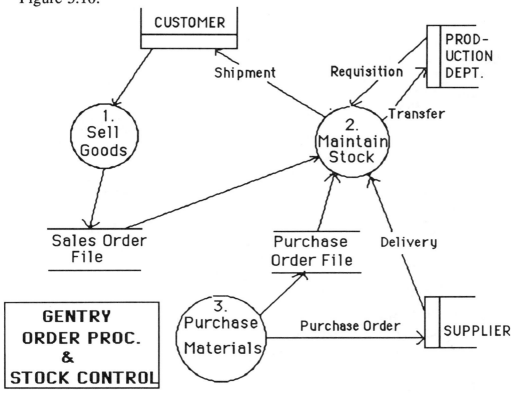

Figure 5.9   Example of a Level 0 Data Flow Diagram

Data stores are normally introduced into the data flow diagram at the level where they are addressed by more than one process. If all access to the store is contained within a process, then it does not need to be shown on the diagram (unless the process is a functional primitive). See, for example, the stores which are recorded in Figure 5.10, but not in Figure 5.9.

When a data store 'emerges' at a particular level, it is expected that information should be both put into it **and** taken from it. Otherwise it would appear that the store was either never updated or never used! The only exception to this is where a data store is **shared** with another system.

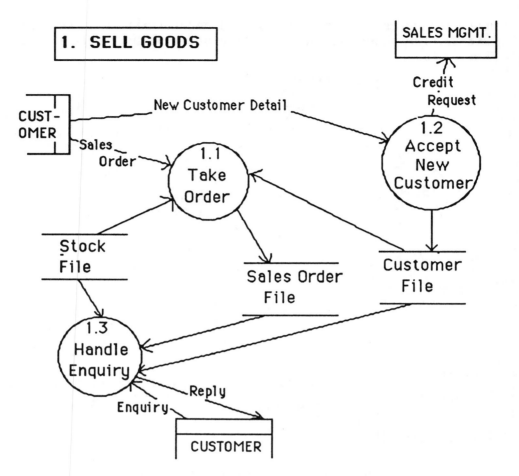

Figure 5.10 Example of a Level 1 DFD

4 CREATING A FULL DATA FLOW DIAGRAM MODEL

4.1 Using the Business Function Diagram

In the Systemscraft methodology, the Business Function Diagram is used to provide the functions or processes for the Data Flow Diagram. The functional decomposition performed in the BFD is also used to indicate the level at which each process or sub-process should occur in the data flow diagram. Figure 5.11 shows how this works.

It should be noted that every page in a DFD model describes one whole function from the Business Function Diagram, and should be accorded the name of the function, (see Figure 5.10). The page itself will include all of the sub-functions which are the children of that 'parent'.

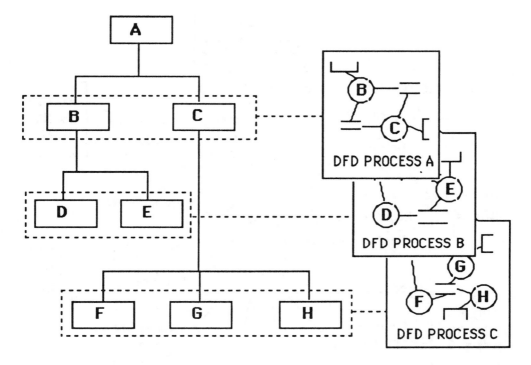

Figure 5.11 Building a Data Flow Diagram from a BFD

4.2 The Context Diagram

In some other methodologies, where the Business Function Diagram is not used, an alternative form of diagram is made use of to initiate the process of building a DFD. This is known as the Context Diagram, and consists of one central process circle (representing the whole system under study), surrounded by and connected to all the external agents of the system. The connections indicate the information which is passed into and out from the system. An example of a Context Diagram is given in Figure 5.12.

The Context Diagram is normally constructed in the first stage of analysis, and is used to help plan the system boundaries, as well as to force the analyst to consider all the system's external references. It should also be obvious that the diagram can be used as the highest possible level of a DFD, in that Level 0 diagram processes can be arrived at by functionally decomposing the Context Diagram's central process.

However, in the Systemscraft methodology it is not considered essential, and the description has only been included in this book for completeness.

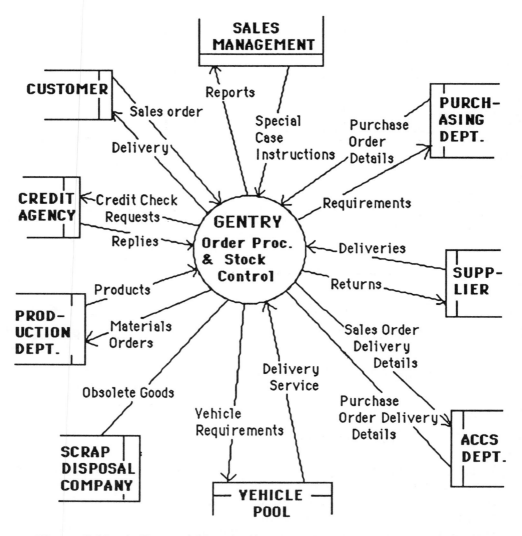

Figure 5.12   A Context Diagram for an 'Order Processing and Stock Control' System

5   THE LOGICAL AND THE PHYSICAL

In the early systems methodologies, where the Data Flow Diagram tool was first made use of, it was considered necessary in the systems development process to create not one DFD model, but four. The thinking behind this related to stages in the systems development cycle.

1. It was felt the existing system in its current physical form should be modelled. The names of actual documents, and the files that were used should be recorded (though the names of the processes were required to be recorded in verb-object form). In particular, the errors and weaknesses of the existing system should be recorded, in order to help the later analysis of the system. This was part of the Investigation stage of the development process.

2. This first model should then be analysed in detail, and all the physical constraints should be removed, These constraints include
>   the people doing the tasks,
>   the location where the tasks were carried out,
>   the time at which the tasks were done.

    This constraint-free version of the system should be modelled, producing a 'logical' DFD. Such a model represents the business requirements inherent in the existing system, and nothing more. As a result, it contains no errors or weaknesses: obviously a weakness can not be a requirement!

3. A third DFD is built using this logical model of the existing system. The extra requirements specified by the system owner are added, completing the requirements specification.

4. The final DFD is again a 'physical' model, unlike the conceptual models of 2 and 3. It depicts how the new system is to work.

An important point which helps to emphasise the distinction between 'physical' and 'logical' in the context of DFD modelling is the fact that it relates directly to the distinction between a 'process' and a 'function'. Strictly speaking, a process is something which can be defined and carried out in physical terms, whereas a function is a purpose, a reason, a requirement, and can be implemented in any number of ways (any one of which could be defined as a process).

So, whereas a circle in a physical DFD can be said to represent a process, a circle in a logical DFD should represent a function. In practice, there is no difference to the approach in naming functions and processes, but it is as well to bear in mind the above distinction when preparing DFDs; successful model-building is very much a matter of attitude.

It has been found in practice that there can be dangers in creating a DFD model of the existing system in its physical form.

1. Firstly, it can take a great deal of time, with users and analysts building, and checking the accuracy of, a detailed model of a 'flawed' system. During this process, no attempt is made to identify or correct the flaws: obviously the intention is to eventually irradicate these weaknesses, but this is to take place in another model done at a later stage. The modelling process can be seen as duplicating the work of other investigation techniques, and therefore consuming project development resource unnecessarily.

2. There is a seductive element about the modelling of the existing physical system. It is a relatively easy process, as the DFD tool is simple, and no hard analytical thinking is required; one is only recording what is done presently, something the user should know extremely well.

3. Once the model is produced (and time is pressing), there is a temptation to make token adjustments to it, and treat it as a logical model, without going through a rigorous analysis process. The eventual result is a new system which is simply a computerisation of the old system, with many of its faults.

The Systemscraft methodology does not incorporate a model of the existing physical system. This does not in any way imply that a full investigation of the existing system should not be done! On the contrary, the investigation details are required for the analysis and checking processes during which models of the logical system are created. It simply suggests that it is rarely necessary or worthwhile to 'model' the existing physical system.

Two situations do however occur where there may be advantage in using the DFD to record how the existing system works. These are:

1. where the existing system is unclear and confusing, there may be advantage in using the DFD informally, to help clarify the situation. The resulting diagrams may be held with the analyst's working papers, but would not be considered part of the Requirements Specification.

2. where the proposed new system will not be ready for some time, and there are aspects of the current system which can be improved in the short term while waiting for the new

system. This situation is not unlike that of a system in the maintenance and development stage, where the current working system must be amended to cater for an extra requirement. In those circumstances however, one would expect that the DFD of the current physical system will already be available.

The Systemscraft methodology endorses both the 'existing logical' and 'required logical' models as essential components of the Analysis stage. It does however recognise that in many situations the 'required' model will not differ significantly from the 'existing': often the users want basically the same system as their present one, only they want it to work better, cheaper, with larger volumes, and in a more flexible way. These are not extra logical requirements, they are physical requirements which must be handled by the proposed design. In such a case, the two logical models of the analysis stage would be identical.

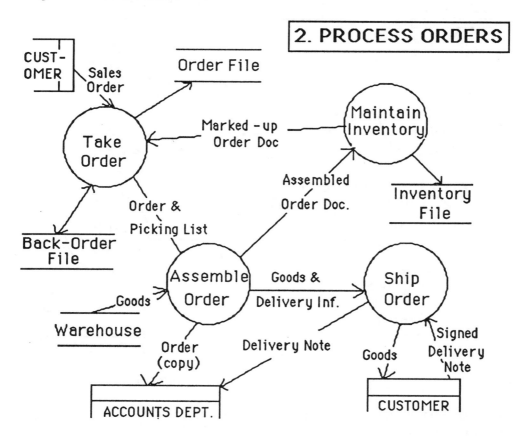

Figure 5.13 Example of a Physical DFD

It is suggested that, in most cases, the required logical model should be built by adjusting the existing logical model, rather than by starting from scratch. After all, only this one final logical version of the model is to be made use of in the Requirements Specification and carried forward into the design stage.

An example of the difference between a physical and logical diagram occurs in Figures 5.13 and 5.14. Both describe the same situation, but whereas one shows the documents involved, the other shows the information necessary. This is particularly noticeable where the physical model illustrates how several copies of the order are sent to different locations, while the logical model shows this as information in a logical store, accessible to all the functions that need it.

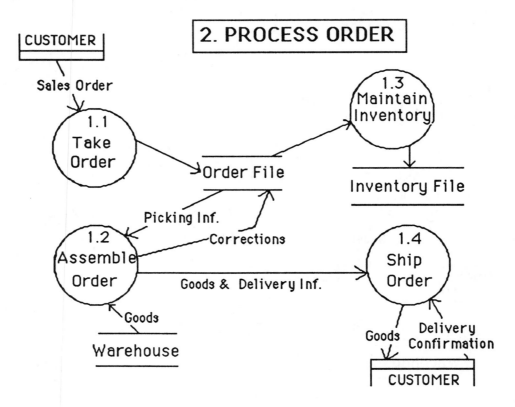

Figure 5.14   Example of a Logical DFD

A second point relating to the logical and physical diagram examples helps to emphasise the difference between information and the documents that carry it. In the physical version (Figure 5.13), a delivery note is shown as being sent to the customer, and then returned to the 'Ship Order' process (a double-arrowed line). However, the information sent to the customer (notification of goods shipped) is different from that being sent back to the process (confirmation and acceptance of receipt). This fact is acknowledged in the logical diagram (Figure 5.14).

One important fact, which will eventually become evident to the modeller, is that there is no clear, unequivocal division between what is 'physical' and what is 'logical'. All models will have some physical elements in them, and the further one decomposes a model the more physical it is bound to become; (at the lower levels one is forced to say 'how' rather than 'what'). Bearing this in mind, the analyst must try to keep the model as logical as possible, as deep into the detail as possible.

The way the methodology handles the construction of the 'required physical system' DFD model(s) is covered in detail in Part 3 of the book (Systems Design).

6 THE NATURE OF INFORMATION

This distinction between the logical and physical models of the system at different stages of the development process should not be confused with the entirely different concept of the 'real world' system with the information system controlling it. This is based on the fact that the 'paper' documents and files which, for the most part, we are using as evidence of information flows and stores, are often merely a system to control the real activities of the company, which involve taking deliveries of materials, moving them about the warehouse, making goods from them, shipping them out to customers, etc.

The problem facing any methodology making use of the DFD tool is whether to model the presence and movements of these goods, payments, materials, parts and maintenance services. Are such movements data flows?: can these physical assets be considered as 'information'?

One of the main purposes of the DFD is to show, for a particular function, all the information necessary to enable that function to be carried out.

1. Some of the information flows may not exist in physical form; for example, some orders may be sent by telephone.

2. Other information may be used directly from the clerk's head; for example, the details of current prices for items on an order.

85

3. Yet more information may be implied by the presence or absence of goods; for example, when the warehouseman goes to pick goods for an order, there may be insufficient items in the bin. This information about inaccuracies in the records, and on the company's need to obtain more stock, is an important 'trigger' for a series of processes.

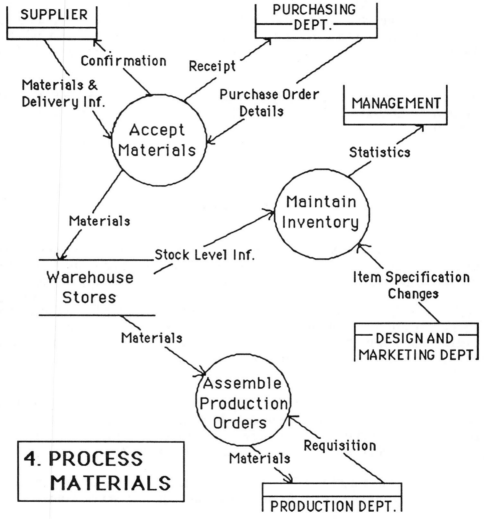

Figure 5.15 A 'Real World' Logical DFD

All three of these examples demonstrate the need for a broad definition of the term 'information'. This methodology encourages the modelling of these movements and presences in the real world, on the grounds that they carry with them implied information. Having said that, some level of judgement as to how much of the real world to include, given the particular circumstances, must be left to the analyst.

Figure 5.15 illustrates a very 'real world' oriented system. Nevertheless, it is still a perfectly valid logical DFD.

7   REVIEWING THE TWO MODELS

When the Data Flow Diagram has been completed, it should be examined carefully for completeness and consistency. Every attempt should be made to make it as simple, as accurate, as logical, and as aesthetically satisfying to both user and analyst as possible.

In particular, three different kinds of situation should be looked for in the model. If they occur, then both the DFD and the Business Function Diagram on which it is based need to be reviewed and revised.

7.1   The 'Sunburst' Effect

The decision as to what is to be a 'process' in the system is originally made during the construction of the Business Function Diagram, when all the 'data' aspects of the system are being ignored. When these processes are used in the construction of the DFD, it sometimes happens that a very large number of data flows relate to one process, to such an extent that the diagram is overcrowded and confusing. This phenomenon is referred to in several of the Yourdon-based methodologies as the 'sunburst' effect, and it can have a number of possible solutions.

It may be, for example, that the process detail is sub-divided at a lower level, and that the flows recorded at this higher level can be grouped and generalised for clarity.

A further possibility is that the BFD was not taken down to a sufficiently detailed level of decomposition, thus producing excessively complex functional primitives. Obviously, the solution is to create that extra level in both diagrams.

It is also important, however, to consider the possibility of re-grouping the components which make up the over-complex process, perhaps distributing some of them among other of the processes on the page (even among processes on other pages). The original decision to group them was made on limited analytical and aesthetic grounds; a new examination may produce an improvement. Often when this happens, the names of the processes need to be changed to reflect their new constituents (and, of course, both models must be brought into line).

7.2   Non-transformation of Data

The analyst must examine each process in the DFD to make sure that it is a genuine process, inasmuch as it TRANSFORMS information. This topic has already been mentioned, but is worth re-statement at this stage, because it can be such an easy mistake to make during the model-building. The occurrence of this error is usually the result of a failure to logicalise the physical situation, and the solution tends to be obvious.

7.3   Processes with No Data in Common

If, while constructing a page of the DFD, the analyst finds that it contains sub-processes which have no data in common (or perhaps only very trivial common data connections), then it is likely that the sub-processes involved are not naturally related, and should not therefore be part of the same higher level process (which the page represents). Almost always, the fault lies with the earlier Business Function Diagram, where functions have been incorrectly decomposed. The components of this higher-level process should be re-examined, and redistributed in a more meaningful way.

The thinking behind this is that each business function of the organisation will make use of specific information to achieve an objective, which itself can be thought of as information. All of its sub-functions must be working towards this common aim. If we find that some of the sub-functions do not contribute to this target information, and apparently do not support or make use of any of the intermediate information products of the function, it is hard to see how the sub-function can genuinely relate to the function.

8   SUMMARY

The Data Flow Diagram is one of the most important tools in structured systems analysis. It provides a method of relating the functions or processes of the system to the information they make use of. It is an essential part of the Systems Requirements Specification, because it defines what information MUST be present before a process can take place.

A DFD can be either 'physical', representing what actually happens (or is to happen), or 'logical', representing functions which need to be carried out (without saying how). In the Business Analysis stage of the development we are primarily concerned with the logical model.

In the Systemscraft methodology the DFD is built using functions identified in the modelling of the Business Function Diagram. On completion of the DFD, both models need to be reviewed for accuracy, consistency and balance.

6 ANALYSING THE BUSINESS INFORMATION

THE ENTITY MODEL

This chapter is an introduction to an entirely different but complementary approach to modelling the business system. The approach is known by several different names, the most common of these being entity modelling, entity-relationship modelling, data modelling, logical data analysis or just data analysis. It is an important and complex topic, incorporating a set of techniques which can be made use of at a number of different points in the development process. Some systems methodologies, for example the early Demarco/Yourdon variants, place little emphasis on this form of analysis, whereas for others, such as CACI and Information Engineering, it is the most dominant of the approaches used. The methodology described in this book, in common with most of the better known commercial methodologies, attempts to balance the use of Data Analysis techniques with the equally important Process Analysis techniques already discussed.

The chapter consists of

1. A discussion on the need for a data-oriented approach

2. A definition of entity modelling and its various components

3. A description of techniques used in building the model

4. A brief illustration of some of the more advanced data modelling techniques.

Later chapters will take the data modelling process into much greater detail, and will illustrate how the two different views of the system (one based on the data and the other based on the processes) are reconciled.

It is best to begin with a broad definition of the topic, the detail of which will gradually emerge through the rest of the chapter.

> **Data Analysis is the analysis of the structure of the information to be used and held within the system under study. This analysis goes beyond the examination of the user requirements; it enables the analyst to identify all the natural relationships between the basic information components. In this way it unlocks and exposes the potential of the information.**

The version of the data modelling technique described in this chapter is one in which an initial model can be built very quickly (eg. within the first two days of the study). This makes it particularly appropriate for use within an Evolutionary Systems Development methodology, in that this model can be used to help build a preliminary database for early prototypes.

1   THE NEED FOR A DATA APPROACH

In the past, systems analysts have tended to consider the data in the system only from the point of view of the functions, processes and programs with which they were concerned. They would address themselves only to the aspects of the information that were directly of use in the task they were specifying, and they would design computer files structured to complete those tasks.

But the information has far greater potential than that! Often, as soon as a computer system is installed, the users begin to ask for modifications and improvements. These changes nearly always involve making use of the same basic information; what happens is that the users begin to see ways of making more effective use of it.

Unfortunately these new user requirements sometimes cannot be accommodated cost-effectively because of the original design approach taken by the analyst. What may seem to be a minor alteration in the user's view may in fact involve a complete re-structuring of several data files, or massive processing overheads to overcome the physical limitations of the original system solution. If this new need had been forseen, an entirely different design approach might have been taken.

In the design of computer files, information is often accumulated and combined with other information which is used in the same processes. But the same information may be of use in entirely different processes within the business. This can mean that some data is duplicated a number of times. For example, it is not uncommon for the name and address of an employee to be held on a number of different files within an organisation (perhaps a payroll file, a personnel file, an 'emergency call' file etc.). When the employee notifies a change of address, there is not merely a danger but a likelihood of one or more of the necessary file amendments being overlooked. So not only is there a waste of space in keeping duplicate information, but also a potential for inconsistency.

Data Analysis is a method of identifying the basic units of information which are of use in the system and the inter-relationships or cross-references between them. This means that any piece of data (eg. an employee address) needs to be held once only throughout the organisation's system, and would be accessible to any number of programs; a place for everything and everything in its place.

2   THE ENTITY MODEL AND ITS COMPONENTS

It has been mentioned that the approach being discussed is sometimes refered to as Logical Data Analysis. Another word that can be used instead of *logical* is *ideal*: data analysis means taking an *ideal* view of the data, considering the best theoretical way it could be held.

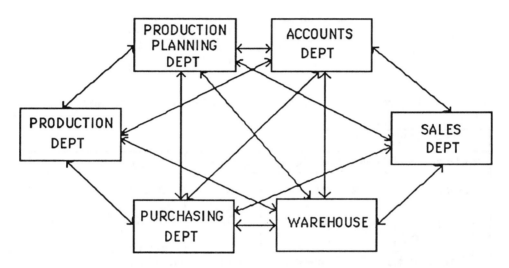

Figure 6.1 Information Transfer within an Organisation

As an illustration of this, Figure 6.1 gives an impression of the flow of information within a fairly standard organisation, showing the complexities of the communications between departments. An ideal approach to this communication problem would be to hold mutual information in common, in some kind of 'information warehouse', much as one would hold stock. Information is after all just as much a resource and an asset of the organisation as is stock! Figure 6.2 gives a revised picture of the organisation's communication system.

So the ideal picture of the system's information is close to that one might expect to find if a good database management system software package is used. This does not mean that the physical implementation of the designed system must incorporate a DBMS, or that it is not worth using Data Analysis techniques unless DBMS software is available. In fact there are major advantages in using data analysis in the logical phase of the project, irrespective of the physical file handler chosen for the implementation.

93

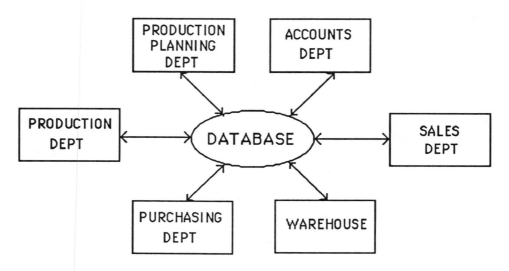

Figure 6.2 The Ideal Method of Information Transfer

CUSTOMER

CustomerNo	Customer Name	Credit	Address
A137	J. SMITH	A	10 HIGH ST BUDE
A257	CATESBY	B	20 LOW ST BATH

ORDER

Order No	Customer No	Date	Delivery Address
2001	A137	5.7.B9	25 TIPTON WICK
		7.7.B9	7 ANDOVER MEWS

ORDER LINE

Order No	Item No	Qty Ordered	Units Ordered
2001	63B	25	GRDSS
2001	47B	14	DOZEN
2003	14C	9	INDIV.
2003	47B	22	DOZEN

Figure 6.3 Examples of Logical Tables and their Possible Contents

94

Assuming that this logical approach is to be taken: How is the detail of the information to be shown? What are to be the shelves and rows of our 'information warehouse' and How are they structured and organised?

The answer is: in the form of **TABLES**. All information in the database can logically be thought of as being held in simple tables, each table having a fixed number of columns (Figure 6.3 gives some examples). The skill involved in data analysis is that of assigning each piece of information to the most appropriate table.

At the analysis and design stage of a system development project, the analyst must identify the kinds of table needed for the data, and must indicate the natural relationships between entries in these different tables. The lozenge from Figure 6.2 must be expanded to the next level of detail, and the tool used for this is the Entity Model.

Figure 6.4 gives an example of such a model. Each box on the chart represents a separate table (the box name describing the nature of each entry or row in the table), and the lines and arrows indicate the relationships.

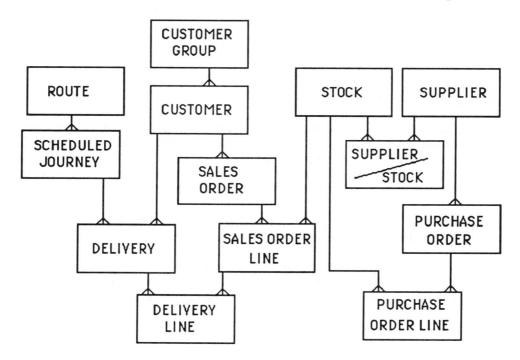

Figure 6.4   An Example of an Entity Model for part of an Organisation

An Entity Model is also known as a Logical Data Structure Model, and is constructed using four types of building block, the following major logical concepts:

1. an Entity
2. an Entity Type
3. an Attribute
4. a Relationship.

2.1   An Entity

An entity is a topic, task, object or event of interest to the organisation (and within the area of the system) and about which information is kept. Some examples of entities are

Mr J Smith   (customer)
Order number   2732547
Stock Item   AB25 (a 5cm spanner).

An entity is the equivalent of an entry (or row) in one of the tables.

2.2   An Entity Type

An entity type is a natural grouping of entities, a description of a kind of information, rather than the information itself. For example, 'Mr J Smith' is an entity, but he is of interest to the Organisation because he purchases things from it; he is a customer of the firm. 'Customer' is an entity type, as it describes a number of entities on which information is to be held.

An entity type is the equivalent of a logical table, and of a box on the entity model chart (in fact, entity types are sometimes refered to as entity tables).

Newcomers to the technique of Data Analysis sometimes experience difficulty in identifying the main entity types needed to begin the modelling process. More will be said later about the approach that the analyst can take, but at this stage it is worth pointing out that the most important entity types fall into one of three categories:

1. Information relating to one of the system's major **TRANSACTIONS**. (eg. sales orders, purchase orders, invoices, etc.).

2. Information relating to the organisation's **ASSETS** or **RESOURCES**. (eg. stock, customer, supplier, staff, materials, etc.).

3. Summarised information, often in the form of statistics, relating to **PLANNING** or **CONTROL**. (eg. forecast, budget, sales performance, vehicle schedule, etc.).

(It is important to appreciate that this categorisation is only helpful at the start of modelling. As the process continues, more intricate entity types begin to surface, and there is no benefit in attempting to categorise them.)

The simplest approach in deciding whether a type of information should be included as an entity type in the model is to ask oneself if a TABLE of such information might be useful to the system, and if so, to identify the basic entity which will constitute an entry or row in the table. For example, a 'stock' table will consist of one line for each type of product held in stock (see Figure 6.4).

2.3   An Attribute

Having identified the appropriate entity types (tables) and the nature of the entities (rows), the next stage is to identify just what information needs to be held for each entity.

Attributes are the characteristics of the entity, represented by the fields or columns of the table. For example, in Figure 6.3, the attributes for a customer (see customer table) are

> the customer number
> the customer name
> the credit rating
> the customer address.

So attributes relate to particular entity types, and specific attribute values belong to individual entities (eg. 'customer' entity type has 'customer number' attribute, whereas 'Mr Smith' has customer number 'A137'). All entity types must have at least one attribute,

There are three different types of attribute, and any particular entity type may have a number of examples of all three. The types are

> 1   Identifiers
> 2   Descriptors
> 3   Connectors

2.3.1 Identifier Attributes

It is essential that each entity in a table can be uniquely identified. There can not be two or more different entities with the same identifier, as it would not be possible to decide which referred to the situation being addressed. On the other hand, if the duplicate entries refer to the same entity, there is no advantage in holding more than one copy.

An IDENTIFIER (also known as a 'key') is one or more attributes in an entity type which are used to give each entity its uniqueness.

For example   (see Figure 6.3)

> **Customer Number** is the identifier for the 'customer' entity type. This means that every customer in the table must have a different customer number. The Customer Name attribute might have been a candidate for identifier, but there is always the possibility of two or more different customers having the same name, in which case there could be duplicate entries in the table.

> **Order Number** is clearly the identifier for the 'order' entity type.

> However the entries in 'order/item' (which is a table containing every item line from every order in the system) are not uniquely identifiable from one attribute. Neither the Order Number nor the Item Number can be used alone; there may be many items from the same order and the same item may be asked for in several orders. The only valid identifier for this entity type is the combination of the two attributes **Order Number** and **Item Number.**

(It may not be clear to the reader who has no previous experience of Data Modelling why it is necessary to treat the item lines of an order as a separate entity type from the order itself. This is explained in more detail later.)

2.3.2   Descriptor Attributes

Most of the attributes in an entity type are likely to be DESCRIPTORS. These are pieces of information which describe the entity being referred to. This information enhances our knowledge of the entity, and will serve a useful purpose within the system.

Examples of descriptors from Figure 6.3 are;

for Customer Entity Type,   Customer Name
                             Credit Rating
                             Address.

for Order Entity Type,   Date
                             Delivery Address.

for Order/Item Entity Type, Quantity Ordered
                             Unit of Order.

The most important thing to note about Descriptor attributes is that each such attribute should occur in one table and one only, and that the art of data analysis is to get the descriptor attribute in the right entity type.

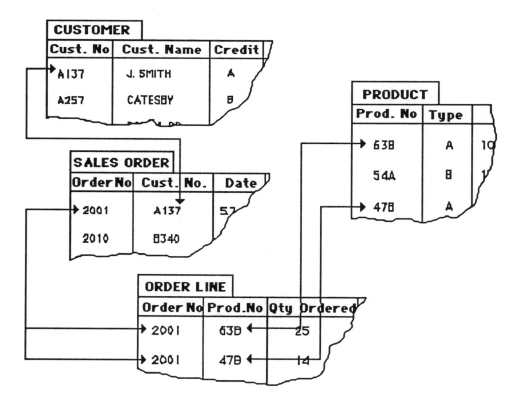

Figure 6.5 Entity Tables highlighting Connector Attributes

99

2.3.3   Connector Attributes

The purpose of a connector attribute is to show the relationship between the entity in which it exists and another entity in another table. The connector attribute is just like an ordinary descriptor attribute in its home entity, but in another entity somewhere else in the model it is an IDENTIFIER. This is best illustrated by example.

In Figure 6.3, the Order table contains the attribute 'customer number'. It is obviously important that we know who has placed the order, so this information must be present. However, the same attribute 'customer number' is also present in the Customer table, only in that table it is the identifier attribute. This indicates the relationship between a customer and an order (an order must be placed by a customer). It also means that there is no need to hold any other customer information (eg. invoice address) in the Order entity table. If we wish to know that address, we can use the customer number from the order to look it up in the Customer table. That is how the connector works.

Figure 6.5 further illustrates the effect of connector attributes in tables. The use of such attributes is integrally linked to our next topic of definition, and the whole concept should become gradually clearer as the chapter progresses.

2.4   A   Relationship

As we have seen, natural relationships occur between entities of different types. One can see for example that relationships exist between

> a customer and an order
> (a customer places orders),

> an order and an order item
> (an order includes order items),

> a product and a supplier
> (a supplier provides products), etc.

Such relationships are essential to the organisation, and are made use of in the conduct of its business.

Relationships are represented on the entity model by lines with arrows or delta signs, (see Figure 6.4).

In some versions of the data analysis process, the description of the relationship is written in text at the side of the line (eg. " belongs to ", "places ", " contains ", etc.). In this version, it is not considered necessary. We are much more concerned with the different TYPES of relationship which can occur between the entities, and which these lines represent.

There are three major types of relationship used in the simplest forms of Entity Model. They are

        1   One-to-One
        2   One-to-Many
        3   Many-to-Many.

These types concern themselves with the NUMBER of entities in a table that are related to one or more entities in another table. They are best defined and described with the help of examples.

2.4.1 The One-to-One Relationship

Assuming two entity tables A and B, a one-to-one relationship exists if

    *   For each entity in table A there is a corresponding entity in table B

AND

    *   Vice versa (ie. for each entry in table B there is a corresponding one in table A).

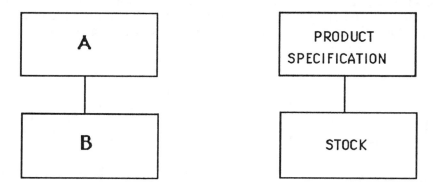

Figure 6.6 Examples of a One-to-One Relationship

In the second example, in a PRODUCT SPECIFICATION entity type, a record (or table entry) exists for each product held in stock, describing its

101

construction and its purpose. This might well be used by the Design or Marketing departments in the company.

At the same time, there exists elsewhere in the system a Stock entity table, keeping details of the quantity of stock being held, the re-order level, etc., and there is obviously again one record in that table for each company product.

Note. The relationship is indicated by a simple line.

2.4.2 The One-to-Many Relationship

Assuming the same two entity tables A and B, a one-to-many relationship exists if

    *   For each entry in entity table A there are MANY entries
        in table B

BUT

    *   For each entry in table B there is ONE and ONE ONLY
        entry in table A.

So there are two directions to a relationship, and in the one-to-many relationship these directions indicate different things.

A one-to-many relationship is shown on the entity model by a line with a delta shape on one end (sometimes called a crowsfoot); the unmarked end indicates the table with the ONE entry, and table at the delta end contains the MANY.

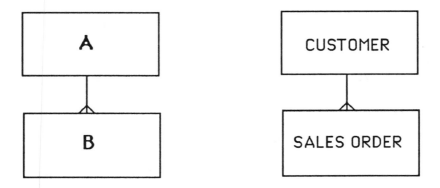

Figure 6.7 Examples of a One-to-Many Relationship

102

In the second example above, the relationship line indicates that for any one customer there may be a number of sales orders, but also that for any one sales order there must be one and only one customer. This makes sense in business terms. It is equally true that any one order may have a number of order lines, but any specific order line must belong to one order.

There is one important point to be made about the use of the word MANY in this context. MANY is taken to mean

zero,
or one,
or more than one.

This enables the model to represent the system as it exists over a full period of time, rather than just to act as a snapshot of one moment in the system's existence.

For example, a customer may normally have a number of sales orders, the details of which are held by the organisation; one may be waiting to be assembled, one may have some outstanding items from last week, and a number of others may refer to orders completed and satisfied during recent months. However, another customer may have just registered with the organisation, and may have not yet placed an order. Similarly, a long established customer may not have placed any orders for a number of months, and the old orders may have been weeded out of the table. The relationship as shown on the model caters for all these situations.

2.4.3   The Many-to-Many Relationship

Using the same two imaginary entity tables A and B, a many-to-many relationship exists if

*   For each entry in table A there are many entries in table B,

AND

*   For each entry in table B there are many entries in table A.

The many-to-many relationship is represented on the model by a line with a delta symbol at each end.

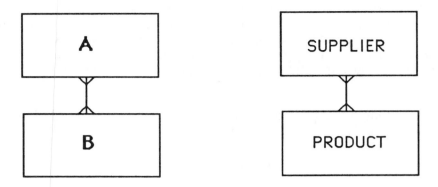

Figure 6.8   Examples of a Many-to-Many Relationship

The second example shown above is a common phenomenon in businesses. Any one supplier is able to provide a number of different products for the organisation to stock. Also any particular product can be obtained from more than one supplier.

The many-to-many relationship is similar to the one-to-many. The difference is that whereas in the one-to-many, the two directions differ (eg. from A to B and from B to A) and one of the entities can be thought of as dominant (the entity at the 'one' end of the model line), in the case of the many-to-many there is no difference in the nature of the directions, and therefore no dominance.

2.5   A Business Rule

The three types of relationship that have been discussed are by no means the only possible types, but they are considered to be the simplest and most common. Of these three, by far the most important is the ONE-TO-MANY relationship.

If a one-to-one relationship occurrence has been identified, as for example in Figure 6.6, there seems little reason to treat it as two separate tables. After all, the entries in the tables have the same identifiers; why should we not simply combine them into one table with longer entries? In fact this is what we are most likely to do in our data modelling process.

The major disadvantage of a many-to-many relationship is that it tells us practically nothing about the business. This is made obvious by the special definition used for the word 'many'. What we are really saying in Figure 6.8 is

For each Supplier there may be zero Products

AND

For each Product there may be zero Suppliers.

We could take two tables from entirely different organisations, and the many-to-many relationship would still be true!

So neither the one-to-one nor the many-to-many relationships are a great deal of use, and you will notice from Figure 6.4 and the other examples in this book that they very rarely occur in developed entity models. Almost all the relationships in these models are one-to-many, to such an extent that another name for the model is a 'one-to-many map'.

We will discuss shortly how a many-to-many relationship can be 'resolved' into a pair of one-to-many relationships, but before that, it is worth emphasising the usefulness in business terms of the one-to-many relationship.

The Data Model is used not only as a file or database design aid, but also as a method of rigorously checking the user's business requirements. This latter property of the model springs from the nature of the one-to-many relationship, and more specifically from the two 'directions' inherent in the relationship.

Whereas the first direction, from the 'one' to the 'many' (eg. a customer can have many orders), is open and non-specific, the second direction, from the 'many' to the 'one' (eg. an order must belong to one customer), is absolutely specific, and represents a 'constraint' which is part of the business requirements specification.

It is recommended in most methodologies that on completion of the Data Model, the relationships should be transcribed into simple text, and their accuracy confirmed with the user.

For example;

' a customer may place many orders over a period of time, but any one order must have been placed by one customer '.

The text description of a relationship is referred to as a 'Business Rule'.

2.6   The Nature of Tables

It has been stressed from the beginning of this chapter that the information belonging to the system has to be considered as being held in logical tables. It is in the nature of tables that they can have a variable number of rows, but each table must have a fixed number of columns. This means that every entity in the table has exactly the same type and number of attributes.

When one first thinks of a Sales Order as an entity type, the natural assumption is that all the information relating to the order will be included in the one table entry. However, the number of line-items is likely to vary considerably from one order to the next, so the amount of information in each order is different. If this information is to be held in tabular form, those attributes (or fields) which are repeated a variable number of times must be identified, and held in a separate entity table. Obviously there would need to be an attribute in this new table to relate these line-item entries to their original order entity (ie. a Connector).

It should be obvious that there are a number of apparent business entities which , like the sales order, when examined closely will be found to have repeated groups of attributes. Examples like the purchase order, the delivery, the invoice, etc. spring immediately to mind. It is necessary for the analyst to examine each proposed entity type very carefully to check whether it is really two or more entity types!

2.7   Resolving Many-to-Many Relationships

One of the likely consequences of not identifying a repeated group of attributes within an entity type is that a many-to-many relationship appears to exist with another entity type.

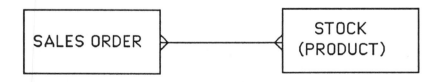

Figure 6.9 A Model where an Entity Type has been Overlooked

Figure 6.9 illustrates the situation where the separate 'order line' has not been identified. As a result there appears to be a many-to-many relationship between 'order' and 'stock'. Clearly any one order might request several different types of stock, and similarly, any one type of stock may be requested on a number of different orders.

Figure 6.10 shows the same situation when the 'order line' entity type has been correctly included. Here the many-to-many relationship between 'order' and 'item' is not relevant. The connection is made via the two one-to-many relationships with the 'order line' table:

1. *An order has many lines, but each line refers to one order,*

2. *A stock item is requested in many different order lines, but each order line refers to one stock item (type).*

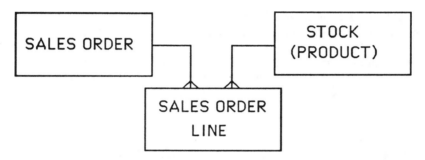

Figure   6.10   The Missing Entity Type Identified

There are however some many-to-many relationships which are not just the result of overlooking a repeating group in one of the entity types. A good example of one such relationship is shown in Figure 6.11.

Figure 6.11   A Genuine Many-to-Many Relationship

In this example, any item in the Product entity table can be supplied by a number of different suppliers, and any supplier in the Supplier table can provide a number of different types of product.

The technique used to resolve such a many-to-many relationship is to create a 'Link Entity Type' which will contain an entry for each actual combination of the entities from the two original tables.

In the example mentioned above, the link entity type might be known as 'Supplier/Product', and would consist of two attributes only; the supplier number and the item number of the product supplied. There will be several entries in the table for each supplier (one for each type of item they can supply) and several entries for each Item type (one for each supplier who can provide it). However, each entry will be a unique combination of a supplier and a product, and these two attributes together will constitute the identifier key.

This construction of a table to show which suppliers provide which products and vice versa may seem contrived, but it is almost exactly the same approach that a clerk would have to take to be able to build a purchase order from a list of requisitions.

The revised data model would now show, instead of the many-to-many relationship, two one-to-many relationships; (see Figure 6.12):

1. *For each Supplier there would be many Supplier/Products, but each Supplier/Product entry would relate to one Supplier only.*

2. *For each Product there would be many Supplier/Products, but each Supplier/Product entry would relate to one Product only.*

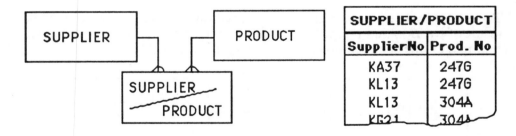

Figure 6.12 The Resolution of the Many-to-Many Relationship

3 BUILDING A DATA MODEL

A number of different approaches can be taken to the building of the initial data model. The one recommended here is one of the simplest in terms of technique, but it is true to say that this approach depends on a good understanding of what constitutes an entity type and a relationship. It allows the experienced data modeller to advance very quickly, but it does not provide, as some approaches do, a series of formal steps to reassure the novice.

There are two elements to the building process:

1. Identifying the entity types,

2. Identifying the relationships.

Although they are described here as occurring in sequence, it is not unusual for the experienced analyst to work on these two elements simultaneously.

3.1 Identifying the Entity Types

This involves identifying the major tables to hold information about the study-area system. It is not critical that all entity types are identified immediately, in fact the model-building process gradually 'winkles out' the more obscure tables as it progresses. So the purpose of the initial stage is to put forward the more obvious candidates for examination and later expansion.

One useful approach in identifying these initial entity types is to make use of the three 'categories' mentioned earlier in the chapter, and to examine the whole study area, posing the question:

Which entity tables are needed to hold details of

1. *the main TRANSACTIONS in the system?*
   (eg. order, delivery etc.)

2. *the main ASSETS and RESOURCES used in the system?*
   (eg. supplies, personnel etc.)

3. *the main information on PLANNING and CONTROL?*
   (eg. schedules, forecasts etc.)

A second and more commonly used approach is to take a precis of the existing or required system, and carefully examine each of the nouns and the more descriptive verbs, asking the question:

> *Is this something on which the organisation needs to hold information, either as an entry in a table or a table in its own right?*

An example can best illustrate this; Figure 6.13 gives a short precis of the Materials Processing system of Gentry Shoes Ltd. The nouns and verbs which have been identified as candidate entity types have been underlined and listed underneath.

> Materials are delivered by the suppliers, and when they arrive, they are checked against a copy of the supplier purchase order. They are then inspected for quality, and, assuming that they are found to be satisfactory, are moved to the appropriate bin for storage. The materials will stay in the warehouse until they are required for production, at which time a production order will be raised, and the requested materials will be picked and dispatched to the appropriate production department.

Materials   Delivery   Supplier   Purchase Order

Production Order   Production Department

Figure 6.13 A Precis of a Proposed System for Study

Some of the nouns in the precis clearly do not form part of an entity table. For example, 'warehouse' does not do so, because there is only one warehouse in the system. Words like 'quality' and 'production' may eventually relate to some entity type, but at this stage their definition is too woolly for us to be sure. It is interesting to note that the verb 'delivered' is identified as a potential entity type; we would base that decision on our general business knowledge, recognising the existence of a 'delivery' transaction in such a system. The other entity types are clear cut; the company must keep tables of all its suppliers, of its purchase orders, of the materials stored, etc.

At this stage some of the most important entity types may still not have been identified, but that should not cause too much concern; almost certainly these entity types will make themselves known as the modelling process continues.

3.2 Identifying the Relationships

Having identified the main entity types, the problem is to identify the natural associations between them, and to record them in the form of one-to-many relationships. The following three statements may serve as guidance in the search for these relationships;

1. A relationship exists between two entities if it is necessary to keep information in one entity about the other. The reason for holding this connector information is the essence of the relationship.

2. In any one-to-many relationship, the entity holding the connector information is, by definition, at the 'many' end.

3. 'Indirect' relationships should be ignored. An indirect relationship is one where the two entities concerned are connected through a third entity. For example a customer may order a particular line of items, but the relationship between Customer and Order Line is not a direct one-to-many; the true relationships are between Customer and Order, then between Order and Order Line.

Some methodologies take the approach of building a matrix whereby all entity types are examined against all other entity types for possible relationships. The approach here is less rigorous and more intuitive.

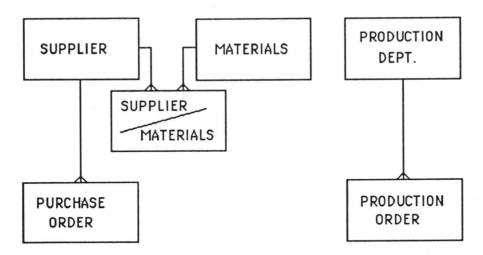

Figure 6.14 First Steps in Applying Relationships

The analyst selects two or three of what are considered to be the most important of the entity types, normally those relating to the ASSETS and RESOURCES criteria (in our example the Supplier, Material and Production Dept.), and tries to identify the relationships between them.

111

Figure 6.14 shows this first step in the process and illustrates how the many-to-many relationship between Supplier and Materials is resolved. It can be seen immediately that one supplier can provide many different materials and any one form of material can be provided by several suppliers. The company must keep a list of which suppliers can provide which materials, so that it can build Purchase Orders, and this list is represented by the Supplier/Materials entity type.

Some of the TRANSACTION criteria entity types, the Purchase Order and the Production Order, are also shown in Figure 6.14. They are relatively easy to associate with their two main sources, the Supplier and Production Dept entity types, by a simple one-to-many relationship. However, how do they relate to Materials?

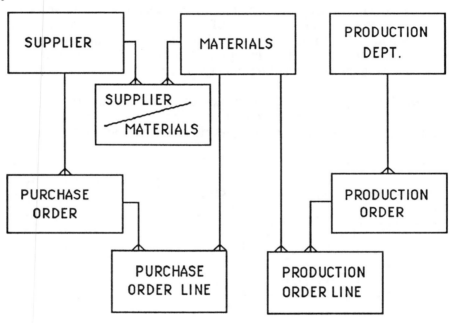

Figure 6.15 Incorporating More Relationships

Clearly there is a many-to-many relationship between the Purchase Order and Material entity types; a purchase order will contain a request for many different types of material, and any type of material will be the subject of many purchase orders over time. This problem is solved by the realisation that a purchase order contains within it a 'repeating group' of attributes, referring to a variable number of individual purchase order lines. The recognition that such a group constitutes a separate entity type enables us to incorporate two one-to-many relationships in the model, one from Purchase Order to Purchase Order Line, and one from Materials to Purchase Order Line.

The same logic applies to the relationship between Production Order and Materials, and Figure 6.15 illustrates both resolutions. (It should be noted that there is a one-to-one relationship between a type of material and a warehouse bin, so those two candidate entity types have been 'merged'.)

There is only one entity type from the original list still to be included in the model, and this can cause some problems. Delivery is clearly related in a one-to-many manner to Supplier, but it is difficult to find a simple relationship with other entities with which it has an affinity. For example, there is usually one delivery for each purchase order, but there are times when a delivery may contain items from a previous purchase order as well. This means there is really a many-to-many relationship between the two entity types.

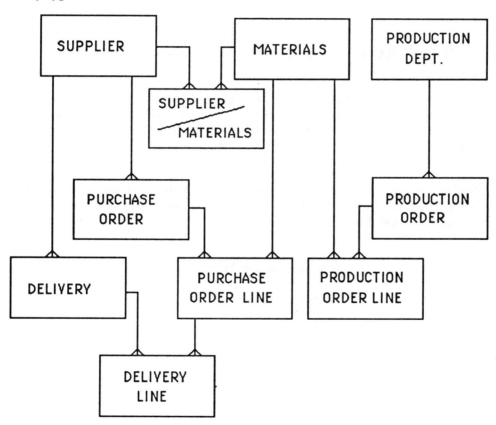

Figure 6.16 The Complete Initial Data Model

On first glance it may seem that this can be resolved by a one-to-many relationship between Delivery and Purchase Order Line. However, again, there can be circumstances where only part of an order line is sent in the first delivery, and the rest follows later, so a many-to-many relationship also exists between these two entity types.

The solution lies in the fact that the Delivery entity type itself has a 'repeating group' of attributes, those relating to a Delivery Line. The connection between the Delivery Line and Purchase Order Line entity types is that a purchase order line may be quoted in more than one delivery (therefore in more than one delivery line), but each delivery line can only refer to one purchase order line. This is shown in Figure 6.16, which completes the initial data model, though more entity types and relationships may well be identified when the attributes are examined.

4 ADVANCED DATA MODELLING CONCEPTS

Although the data modelling concepts discussed so far are sufficient to deal with most simple situations, the more one probes into the detailed aspects of an organisation the more one appreciates the variety of types of circumstance to be modelled, and the need for ways of indicating them on the model.

Four such types of circumstance which cannot be handled by the techniques so far discussed are described here, along with the extra techniques necessary to model them. They are

1 Optional relationships

2 Abstracted data types

3 Wholly enclosed data types

4 Recursive relationships.

4.1   Optional Relationships

The main stalwart of the data model is the essential 'one-to-many' relationship. As has already been stressed, the other types of relationship are normally resolved away, either by merging (as in the case of one-to-one) or creating link entity types (for many-to-many). However, there is an important variation on the standard one-to-many situation, which needs to be recognised, identified and recorded. This is referred to as the zero-or-one-to-many, or 'optional', relationship.

In the normal relationship between two entity types, for example between 'order' and 'order line', it is assumed that the particular entity involved at the 'one' end must be present all the time. For example, it would be nonsense to have a particular order item present in the 'order line' table without having the order to which it related present in the 'order' entity type. This situation is true of the vast majority of such relationships identified.

There are however some situations where at least one entity from the 'many' end can arrive first, and the related entity from the 'one' end may not become available until some time after. For example, in the entity type Delivery Run, each entity represents a day of deliveries made by one driver in one vehicle. However, although the run schedule is set up two days in advance, the particular vehicle is not allocated until a few minutes before the run starts.

It is also possible to have a relationship where there may sometimes be no 'one' entity for a particular 'many' occurrence. An example of this can be seen in the Personnel part of the Gentry system, where each employee may belong to one trade union, but some employees may not be union members.

Both of these situations, illustrated in Figure 6.17, represent 'zero-or-one-to-many' relationships. Such relationships are indicated by a zero symbol at the 'one' end of the relationship line.

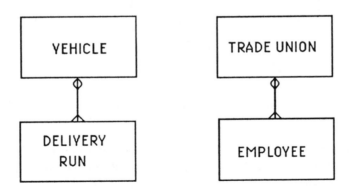

Figure 6.17   'Zero-or-One-to-Many' Relationships

The zero-or-one-to-many relationship is much weaker than its one-to-many counterpart, in that it lacks the 'constraint' element which is so helpful in business requirement specification. It is therefore important that such a relationship is not mistaken for a genuine one-to-many, as this could lead to design errors.

4.2 Abstracted Data Types

Sometimes in a model one identifies a number of different entity types which have some similar attributes and several relationships in common. An example of this is shown in Figure 6.18a, where three entity types representing adjustments to the company's stock position are shown. When this happens, there may be benefits in ignoring the differences between the original entity types, and 'abstracting' them to a higher level.

In these circumstances, 'abstracting' means looking for a more general entity type which could be used in place of the previously discussed entity types which shared the many relationships. This might simplify the overall model to an important extent. Figure 6.18b shows how these three entity types may be combined into one, which might be called Inventory Movement.

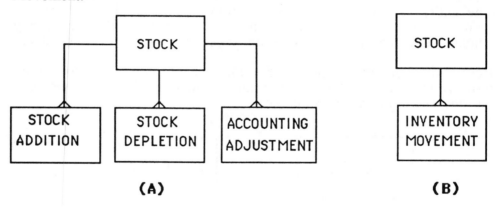

Figure 6.18   Model with 'Abstracted' Entity Type Alternative

It is important to realise that both versions of the model are valid, but one is a clearer and simpler expression of the requirements. The decision whether to go for the 'abstracted' entity type would involve balancing the new clarity of the model against the fact that some of the combined attributes in the new entity type would not have meaning for all cases. (For example, some attributes applying to the Accounting Adjustment would need to be left blank for normal stock additions.)

116

4.3 Enclosed Data Types

This refers to a similar situation to that of the abstracted entity type. Sometimes when building a model there are advantages in describing both the more general entity type and the lower-level entity types of which it is an abstraction together on the same model. This means that there are some entity types on the model which can be wholly subsumed into another entity type which itself is shown on the model.

Figure 6.19 illustrates an example situation where the three different types of stock adjustment are modelled as separate tables. However, some benefit is seen in holding a further entity type, which represents all of the adjustments. As in the previous example, this may be called 'Inventory Movement'. Some of the attributes which were common to all three types would be held in the newly formed Inventory Movement table, whereas those attributes unique to the individual transaction types would be held in their appropriate table.

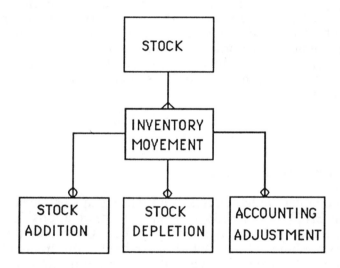

Figure 6.19 An Alternative Version of the Inventory
Model

In this model, there is one entry in the Inventory Movement table for each of the entries in the three stock adjustment adjustment tables.

The type of relationship between each of the original transaction entity types and the new 'Inventory Movement' is interesting. It can be classed as zero-or-one-to-one. This means that

* For each entity in table A there is ONE and ONE ONLY entry in table B

BUT

* For each entry in table B there is ZERO or ONE entries in table A.

This kind of structure can be beneficial in two main ways.

Firstly it can reduce the provision of space for attributes for entities where there will never be a value for them; eg. if there was only to be one entity type, 'Inventory Movement', then NONE of its entities would make use of both attributes which must be present, the Goods-In number (for additions) and the Shipment number (for the depletions).

Secondly it can provide information about all the entities, information that the separate individual entity types are not capable of holding. For example, the 'Inventory Movement' entity type indicates the order in which the various transactions have been applied to the inventory, thereby providing a more thorough audit trail than would be possible if the entity type had not been abstracted.

4.4   Recursive Relationships

So far in this chapter we have discussed relationships between entities OF DIFFERENT ENTITY TYPES; for example, the relationship between a particular customer and the sales orders placed. Obviously the customer and the orders are both entities, but they are held in different entity tables, and it is between the two entity types that the relationship is recorded on the data model.

There is however the possibility of a relationship occurring between entities OF THE SAME ENTITY TYPE. This means that the relationship exists between two or more entries in the same table. This type of relationship is called *recursive*, and obviously a different kind of symbol is needed to represent it.

There are in fact two types of recursive relationship that can exist, the one-to-many and the many-to-many, and each of these is discussed in turn.

4.4.1 The One-to-Many Recursive Relationship

This is the situation where an entity can have many related subordinate entities of the same type, but each subordinate can have only one entity to which it answers within the relationship.

Figure 6.20 Recursive One-to-Many Relationships

Figure 6.20 shows the symbol used to indicate the relationship; (this is often referred to as a 'pig's ear'). It also gives an example of the 'employee' entity type, which very often contains a recursive relationship. The attributes for an employee may include the identity of that employee's line manager. However, the manager is also an employee, and therefore is uniquely identified by an employee number.

The relationship is one-to-many because although each employee will only have one manager, any particular manager may have many staff.

The relationship is recursive because managers themselves have managers, who in turn have managers, etc.

The relationship is supported by the presence of a connector attribute which is of the same type as the entity's identifier.

4.4.2 The Many-to-Many Recursive Relationship

This kind of relationship is relatively rare. It means that each entity in a table can be related to a number of subordinate entities in the same table, and that each subordinate can have a number of entities to which it must answer within the same relationship.

The most common manifestation of this is in a data model dealing with component parts, perhaps in a warehouse or engineering environment. It is often referred to as the 'Bill of Materials', or BOMP situation.

A Motor Vehicle Spare Parts system example might best illustrate this.

119

Basically a part, like say a piston assembly, may have a number of components, one of which might be a particular type of screw. The screw is itself a component of a number of parts, not only of the piston. It may for example also be used in the construction of a radiator.

The recursive element comes in the fact that the higher-level part (piston or radiator) may itself be fitted into a number of even higher-level parts (engine or chassis) etc.

Earlier in the chapter, the nature of an ordinary many-to-many relationship was discussed, and it was pointed out that the relationship could not be supported by the attributes in the table. In order to support a 'many' relationship, there would have to be a variable number of attributes for each entity. This is not possible in a proper data model.

This means that the relationship must be resolved into two one-to-many relationships, and this is done in exactly the same way as with a non-recursive many-to-many, ie. by using a new 'link' entity type.

Figure 6.21 shows how this is recorded on the model. The link entity is called 'Composition/Component' because it will contain two key attributes, both of them part numbers:

the part which contains the component
and
the part which is the component.

Each entry in the link table refers to a unique combination of the two parts, even though each part can occur a number of times as the construction, and a number of times as the component.

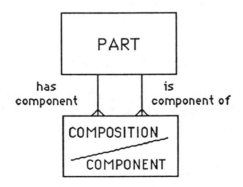

Figure 6.21 Modelling a Recursive Many-to-Many Relationship

The two one-to many relationships are named, and both are fully supported by the link entity type.

One is able to find out all the components needed for a particular construction by looking for all the entries in the link table which refer to that construction.

One is able to find out which constructions require a particular component by looking in the table for all entries which refer to that component.

Figure 6.22 shows a sample of the contents of the two tables representing the entity types in Figure 6.21.

PART			
Part No.	Description	Quantity	Re-order Level
A570	Piston Assembly	50	30
A750	Radiator	48	25
A784	All-purpose Screw		
B213	Engine		

COMPOSIT/COMPONENT	
Composit Part No	Component Part No
A570	A784
A750	A784
B213	A570

Figure 6.22 Tables Supporting a Recursive Many-to-Many
Relationship

5   SUMMARY

This chapter is meant as a brief introduction to the art (or craft) of data modelling. The main concepts have been discussed, and a relatively simple method of approaching the modelling task has been presented. Also, a brief glance has been taken at the more advanced techniques which may be required when constructing more complex models.

The importance of the speed with which the entity model can be built is a key aspect in the evolutionary development process. Many of the modern Fourth Generation Languages used for building evolutionary prototypes are based on some form of relational DBMS, and the entity modelling technique is particularly appropriate in relational database design.

The next chapter will take the data view of the system requirements into much greater detail. Later chapters in the Systems Design part of the book will address the task of proving the completeness of the model, and the techniques for turning this logical view of the information required by the system into an optimal physical database design.

7 CONFIRMING THE BUSINESS INFORMATION STRUCTURE

THE RELATIONAL MODEL

This form of modelling is again part of an overall approach to data analysis. In this methodology it is used as a continuation of the Data Modelling process, to check, improve and extend the already constructed entity model.

A relational model is simply a list of all the appropriate attributes for each of the entity tables of the data model. The order of the attributes within the entity table is immaterial, but those attributes which constitute the unique identifier are underlined. Figure 7.1 shows an excerpt from an example relational model.

The process of creating a relational model and using it to check the data model involves carrying out a number of steps, each of which will be discussed in detail in this chapter:

1. Identifying all the ATTRIBUTES that are to be used within the proposed system.

2. Identifying the appropriate ENTITY TYPE into which each of the attributes should be placed in order to minimise duplication and avoid redundancy. (The technique used for this is known as NORMALISATION.)

3. Identifying the RELATIONSHIPS that are implicit within the attribute lists established for each entity type. This is done by noting which of the attributes in each table are CONNECTORS. Every connector attribute represents the 'many' end of a 'one-to-many' relationship.

4. Given the attributes, entity types and relationships, it is now possible to build a diagram of the same type as the intuitive Entity Model, but based purely on the relational model. We are then able to COMPARE THE TWO DIAGRAMS, and extract from the comparison a single Entity model which contains the best features of both.

5. Having decided on the best format of the Data Model to represent the business requirements of the system, estimates

can be made of ENTITY VOLUMES for each table, and these can be recorded on the model itself.

Although the process of carrying out these steps may consume time and other resources, the benefits of a rigorous approach often outweigh the costs. On the other hand there are clearly times when the system is simple enough not to require the 'full treatment'. Analysts must use their own judgement in such cases.

```
CUSTOMER
Customer Number, Customer Name, Address,
Credit Status

ORDER
Order Number, Customer Number, Order Date

ORDER LINE
Order Number, Item Number, Quantity, Units

ITEM
Item Number, Quantity-in-stock, Re-order Level

DELIVERY
Delivery Number, Customer Number, Delivery Date,
Route

DELIVERY LINE
Delivery Number, Order Number, Item Number,
Qty Delivered
```

Figure 7.1 Excerpt from a Relational Model

1 IDENTIFYING ATTRIBUTES

There are three basic sources from which the analyst gets details of the potential 'attributes' for the entities of the system:

124

1. They can be 'guessed', or identified intuitively, based on the analyst's knowledge of general business practice in the study field,

2. They can be obtained from various users during the interviewing process; (eg. 'What information does the organisation keep about suppliers?')

3. They can be extracted as a result of scrutinising forms and other documents used in the study area.

So, for each entity type on the data model, the analyst builds up a list of candidate attributes. Some attributes may be considered as belonging to more than one of the entity types, in which case, at this stage, they should be included in all the appropriate lists; the problem of this duplication is examined in the next stage of the process.

There is a simple set of standards employed for the recording of these lists, which together constitute the relational model.

* The entity type name is recorded (in capitals),

* The attributes are listed in any order, in a line beneath, each separated from the others by a comma.

* The key identifier attribute(s) are underlined; (normally these are placed at the start of the list for convenience).

It is also good practice to indicate the 'connector' attributes (ie. non-key attributes which are identifiers in other entity tables). These are sometimes referred to as 'foreign keys', and are shown on the model by using a dotted underline.

2 IDENTIFYING THE ENTITY TYPES: NORMALISATION

Normalisation is the name given to the process of examining these lists of attributes, and applying a set of analysis rules to them, converting them into a format which

> minimises duplication,
> avoids redundancy,
> identifies and resolves ambiguity.

Duplication, the situation where the same attribute is in more than one entity table, should only occur for identifier and connector attributes, and is necessary in order to support relationships.

Redundancy can occur in the form of 'derived' data. Attributes which have values resulting from a simple calculation performed on other attributes should be excluded from the model.

An example of such an attribute is 'Cost' in the list

Item Number, Quantity, Price, Cost

(where 'Cost' is simply 'Quantity' multiplied by 'Price').

Not only is it a waste of space to hold this derivable attribute, but it also leaves the database open to the possibility of inconsistency. For example, if an entry in the table has values of

Quantity   25
Price   4
Cost   120

then the entity itself has a built-in error.

Ambiguity can occur in any situation where information is used by more than one person. It can be in the form of an attribute not clearly understood, or one which means different things to different people. The process of normalisation forces a very careful examination of the meaning of each of the attributes, and the relational model cannot be built unless the attributes involved are clearly understood. The analyst is prompted into asking questions like

Is 'customer number' the same as 'customer account number'?

Is there a relationship between 'product type' and 'inventory class'?

etc.

2.1   Functional Dependency

The process of Normalisation is based on the concept of 'functional dependency', and the ideal, fully normalised model is one where every attribute in every entity table has a direct functional dependence on the key attribute(s) of the table.

Functional dependency means that

> *for any one value of a key, at any one given moment, there can only be one value for each other attribute in the table.*

For example, in the entity table CUSTOMER,

Customer Number, Customer Name, Customer Address

the customer may change address, but at any one moment, given the key (Customer number), it could be said that there was one value for name and address.

If an attribute is not functionally dependent on the key (according to our definition), then it should be in another entity table.

For example, in the entity table ORDER,

Order Number, Order Date, Customer Number, Order Item, etc.

one cannot say that, given the key (Order Number), there is only one value for the Order Item attribute; common business knowledge tells us that an order may have many order items.

As the Order Item attribute is not functionally dependent on the key, it is not a valid attribute for this entity type, and must be held in another.

2.2 The Three Normal Forms

The list of attributes for a particular entity table is described as 'un-normalised' until it has gone through the normalisation process. The process of NORMALISATION involves applying three rule-checks one after the other.

* If the entity type list passes the first rule-check, it is said to be in 'first normal form', (1NF).

* If it passes the second, it is said to be in 'second normal form', (2NF).

* If it passes the third, it is said to be in 3NF, and, to all intents and purposes, it is then considered to be 'fully normalised'.

(It must be stressed that there are a number of more advanced 'normal forms', but they are not commonly used in business systems analysis.)

If an entity type list fails to pass a particular check, it means that one or more of the attributes within it must be removed, and made part of a different entity type. This means that an analyst will start with a list of supposed attributes for one entity type, and as the three normal form rules are applied, some of those attributes will be assigned to other entity types. By the end of the normalising operation, not only the original entity type, but also all of the newly identified entity types should be fully normalised.

2.2.1 Choosing a Key

The first task, given a list of attributes for an entity type, is to choose a key. This will consist of one or more attributes whose value will provide a unique identifier for each entry in the table: no two entities in an entity type can have the same key value.

Very often the choice is a straightforward one, but sometimes it requires some thought. Some attributes seem at first glance to offer uniqueness, but after reconsideration can be seen to run the risk of duplication.

For example,

1. in the entity table CUSTOMER,

   'Customer Name' may seem to be a promising candidate key. However, it is obvious that in most organisations there is the possibility of having two customers with the same name, so the customer name is unlikely to be worthy of consideration as a key.

2. in the entity table SUPPLIER ITEM,

   the Entity table describing the different items provided by suppliers contains information needed to build a purchase order. One of the attributes for this table is the 'Supplier's Item Number', identifying the article as an entry in the supplier's catalogue. This number by itself is not a candidate key, because it is possible that two different suppliers could use the same numbering system to catalogue entirely different items. It is in fact rare to find a candidate key from attributes whose values are beyond our control (eg. in the hands of external agents).

Quite commonly, the only possible candidate key is a combination of more than one attribute.

For example,

1. in the entity type SUPPLIER ITEM,

   the combination of 'Supplier Number' and 'Supplier Item Number' would serve as key.

2. in the entity type SALES ORDER ITEM,

   the combination of 'Order Number' and 'Order Item Number' might be the best possible key.

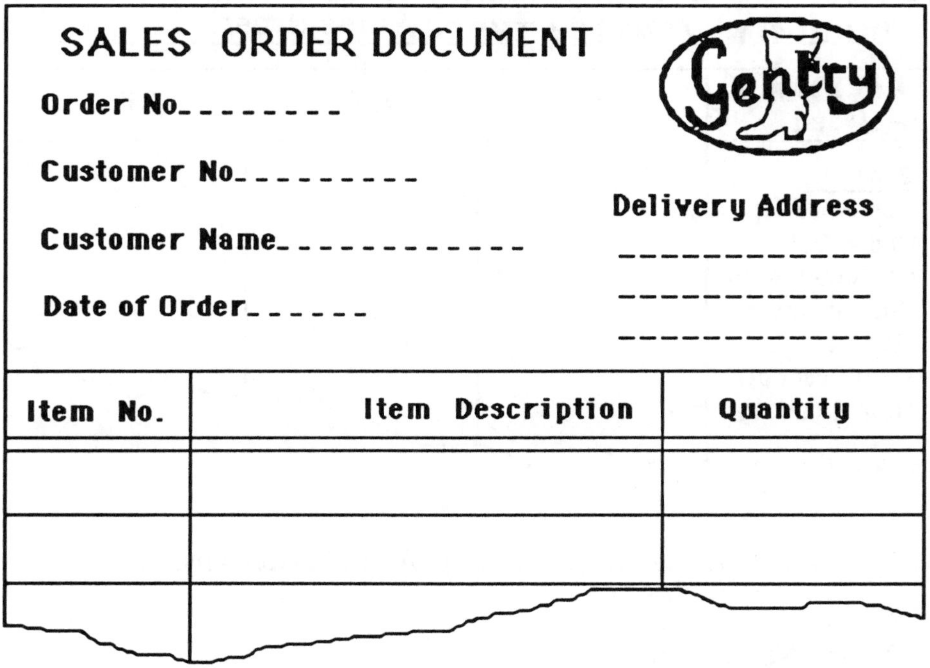
SALES ORDER DOCUMENT

Gentry

Order No.________

Customer No.__________

Customer Name.____________

Date of Order.______

Delivery Address

____

____

____

Item No.	Item Description	Quantity

Figure 7.2 A Standard Order Document

Generally when there is a choice of possible keys, the rule is to choose the smallest, and show preference for numeric rather than alphabetic candidates.

As the relational model is a 'logical' model of the data available to a system, no processing considerations are taken into account when choosing the key. If there is no unique key available in the attribute list, it may be that this is not a genuine entity table. In the worst instance, it may be necessary to design a unique key for the entity type, but this should not be done lightly, as there can be major processing repercussions. The skills required for code design are outside the scope of this book.

In order to illustrate the normalisation process it is useful to take an example of a common business transaction. Figure 7.2 illustrates a simple order form, which might have been used by the analyst to help make a starting list of attributes for the entity type 'Order'.

Figure 7.3 illustrates a particular form of layout which can be used effectively to guide the analyst through the process of normalisation, and its use will be demonstrated as the chapter progresses. It can be seen that the attributes identified from the Figure 7.2 order form have been entered as 'un-normalised', and that the 'Order Number' has been identified as the unique key.

DOCUMENT/ENTITY TYPE Sales Order			
Un-Norm -alised	1NF	2NF	3NF
Order No Customer No Order Date Customer Name Cust. Address Item No Item Descrptn Item Quantity			

Figure 7.3   Form of Layout to Assist in Normalisation

2.2.2 First Normal Form

The first rule or constraint applied to the entity type list is that it should contain no attributes which can occur more than once for the same entity.

For example,

1. in the entity type ORDER,

   all attributes relating to an order line could occur a number of times.

2. in the entity type SUPPLIER,

   all attributes relating to an item supplied by the supplier could occur a number of times (a supplier can supply many different items).

In functional dependency terms, it should be obvious that if an attribute is repeated several times within the entity table, it cannot be said that

for any one value of the (entity) key, at any given moment, there can only be one value for each other attribute in the table.

So the approach required to put the entity type list in 1NF is

REMOVE ALL REPEATING GROUPS

In the example of the standard order entity type list, Figure 7.3 shows all the attributes, and Figure 7.4 shows in the 1NF column the five attributes which only occur once in the entity, the others having been dropped.

DOCUMENT/ENTITY TYPE __Sales Order___			
Un-Norm -alised	1NF	2NF	3NF
Order No Customer No Order Date Customer Name Cust. Address Item No Item Descrptn Item Quantity	Order No Customer No Order Date Customer Name Cust. Address Order No Item No Item Descrptn Item Quantity		

Figure 7.4 First Normal Form for the Order Entity Type

The dropped attributes obviously relate to a different entity type, which itself may be considered for normalisation. Using the approach recommended in this methodology,

    1. the dropped attributes,

PLUS

    2. the key attribute(s) from the entity type of which these attributes are considered a repeating group,

are combined and listed as the new entity type, also in Figure 7.4.

The key from the higher-level entity table is carried forward into this new entity table because there is clearly a 'one-to-many' relationship between the two entity types, and it must be supported by a 'connector'.

The new entity type is then examined for its own unique identifier, and the best candidate is chosen and underlined. It must be stressed that this new key does not necessarily include the key from the higher-level entity type (though very commonly it will). Again, Figure 7.4 shows the chosen key underlined.

2.2.3 Second Normal Form

The second rule applied to each entity type list is that none of the attributes should be dependent on only part of the key instead of the whole key. Obviously, the only entity types that are in danger of not complying with this rule are those with compound attribute keys; any table with a single attribute as key can automatically be considered to be in 2NF.

An example of an entity type list not in 2NF is shown below:

SUPPLIER ITEM

Supplier Number, Supplier Item Number, Our Item Number, Supplier Address, Contract Identifier

In this example, the attribute 'Our Item Number' is dependent on the whole of the key; it would make no sense to say that if we knew the Supplier Item Number we could quote Our Item number, because it has already been seen that different suppliers may use the same number to represent different items. We need to know both the Supplier Number and the Supplier Item Number. So the attribute is dependent on the combined key.

However, the attribute 'Supplier Address' is dependent, not on the whole key, but only on the Supplier Number part of the key. It can be seen that irrespective of the value of the Supplier Item Number, a particular supplier (and therefore a Supplier Number) will have one address. This indicates that the attribute Supplier Address should be in a different entity type (SUPPLIER).

But what of the third attribute? Does the contract relate solely to the supplier? Can there be a different contract for each item supplied? The answer to this depends on how the organisation handles its business, and it is worth noting that the normalisation process forces the analyst to probe deeply into the business rules and methods in order to identify the true meaning of attributes.

So, the approach required to put an entity type list in 2NF is

REMOVE PART-KEY DEPENDENCIES

DOCUMENT/ENTITY TYPE __Sales Order___			
Un–Norm –alised	**1NF**	**2NF**	**3NF**
Order No Customer No Order Date Customer Name Cust. Address Item No Item Descrptn Item Quantity	Order No Customer No Order Date Customer Name Cust. Address	⟶	
	Order No Item No Item Descrptn Item Quantity	Order No Item No Item Quantity	
		Item No Item Descrptn	

Figure 7.5   The Three Entity Types in 2NF

133

In the example of the standard order entity type list, the attribute 'Item Description' is only dependent on the Item Number part of the ORDER ITEM key, so it must be left out of its entity type list, (see Figure 7.5).

'Item Description' is however a valid attribute, and belongs to some entity type! Clearly if it is functionally dependent on the Item Number, it belongs in an entity type of which Item Number is the key.

So a further entity type , ITEM, is introduced on the normalisation form. The Item Number attribute is identified as its key, and it is naturally already in 2NF, (see Figure 7.5).

(It should be noted that the method used to indicate on the normalisation document that an entity type list does not need to be altered, but is already in the higher normal form, is to show an arrow passing it through the NF column.)

2.2.4 Third Normal Form

The third normalising rule states that not only must each attribute be functionally dependent on the key, it must also not be functionally dependent on any other attribute in the list.

The situation can occur where an attribute appears to be functionally dependent on the key,

For example,   in the entity type SUPPLIER ITEM

<u>Supplier Number, Supplier Item Number</u>, Our Item Number,
Supplier Item Description, Our Item Description

the attribute 'Our Item Description' is clearly dependent on the whole key (ie. for every supplier item, at any given moment there will only be one description used by our organisation).

However, the 'Our Item Description' attribute can also be seen to be dependent on the non-key attribute 'Our Item Number', and that dependency is considered to be the more significant of the two. When there is a non-key attribute dependency like this, the other dependency on the main key is referred to as 'indirect', using the logic:

C is only dependent on **A**
because **B** is dependent on **A**
and C is dependent on **B**.

134

So, the approach to be taken in order that an entity type list will be in 3NF is

REMOVE NON-KEY DEPENDENCIES

The attributes which have the dependency on the non-key attribute must be separated from the entity type list, and set up as a new entity type, the key of which will be that same non-key attribute.

In the Supplier Item example discussed above,

1. the SUPPLIER ITEM entity in 3NF would look like this;

    <u>Supplier Number, Supplier Item Number</u>, Our Item Number, Supplier Item Description

2. the new OUR ITEM entity would be in 3NF, and look like this;

    <u>Our Item Number</u>, Our Item Description

(Note that the new key 'Our Item Number' is also retained in the original entity table. This is important, because it acts as a foreign key, supporting a one-to-many relationship.)

DOCUMENT/ENTITY TYPE __Sales Order___			
Un−Norm−alised	**1NF**	**2NF**	**3NF**
<u>Order No</u> Customer No Order Date Customer Name Cust. Address Item No Item Descrptn Item Quantity	<u>Order No</u> Customer No Order Date Customer Name Cust. Address <u>Order No</u> <u>Item No</u> Item Descrptn Item Quantity	⟶ <u>Order No</u> <u>Item No</u> Item Quantity Item No Item Descrptn	<u>Order No</u> Customer No Order Date <u>Customer No</u> Customer Name Cust. Address ⟶ ⟶

Figure 7.6 The Four Entity Types in 3NF

135

Figure 7.6 illustrates an occurrence of a non-key dependency, where 'Customer Name' and 'Customer Address' are functionally dependent on the 'Customer Number' attribute. The Figure shows how the required actions are recorded on the normalisation form.

It can be seen from Figure 7.6 that, having started with a list of attributes apparently relating to one entity type (ie. ORDER), we have now identified four entity types with which different attributes from the list should be associated (ORDER, ORDER ITEM, ITEM and CUSTOMER). The resulting relational model, shown in Figure 7.7, is said to be fully normalised.

```
ORDER
Order Number, Customer Number, Order Date

CUSTOMER
Customer Number, Customer Name, Address

ORDER LINE
Order Number, Item Number, Quantity

ITEM
Item Number, Item Description
```

Figure 7.7 The Relational Model from Figure 7.6

2.2.5 Merging Common Entity Types

Very commonly, the same entity type will be identified from several of the entity table lists when they are normalised.

For example,

a CUSTOMER entity type has been identified from the standard Sales Order, it is quite possible that a CUSTOMER entity type may also have been identified from an examination of say the Customer Registration.

We would then be in the position of having two identical entity types, both in third normal form, but each perhaps having some different attributes.

1. Entity Type CUSTOMER (from standard order)

   <u>Customer Number</u>, Customer Name, Customer Address

2. Entity Type CUSTOMER (from customer registration)

   <u>Customer Number</u>, Customer Name, Credit Limit, Status

The different versions of the entity table must be MERGED so that all relevant attributes are associated with the one entity type.

The attributes 'Number' and 'Name' are present in both versions, and must be carried through to the merged entity type, and so must the other attributes which are present in only one of the versions. The final merged version of the CUSTOMER entity table list should be as follows:

   <u>Customer Number</u>, Customer Name, Customer Address,
   Credit Limit, Status

There is a possibility that when two or more versions of an entity type attribute list are merged, even though each version separately is in 3NF, the resulting merged version may only be in 2NF. This should be obvious from the fact that the merged version may contain a combination of non-key attributes which have never before been considered together; they could contain a non-key dependence.

Therefore, after the merge process has been carried out, the test for 3NF must be re-applied.

An example will help to illustrate this.

There may be two versions of the entity type ORDER,

1. ORDER (from the order document),

   <u>Order Number</u>, Customer Number, Date

2. ORDER   (from the delivery document),

    <u>Order Number</u>, Order Status, Delivery Address

In the merged entity type list, it may be found after examination that there is only one delivery address for each customer. In that case, although both original order lists were in 3NF, the merged list is in 2NF, as it contains a non-key dependency.

The merged entity type list,

    <u>Order Number</u>, Customer Number, Date, Order Status, Delivery Address

should be revised to show;

    ORDER

    <u>Order Number</u>, Customer Number, Date, Order Status

    CUSTOMER

    <u>Customer Number</u>, Delivery Address

(Almost certainly, CUSTOMER will then merge with further versions of its entity type.)

2.2.6 Summary of the Normalisation Process

The process of building a relational model based on the entity types of the study area consists of the following stages:

1. Listing the un-normalised attributes identified for each entity type.

2. Applying the three NF rules and producing fully normalised relations (entity type lists).

3. Merging the different versions of the same entity type lists.

4. Re-applying the 3rd NF rule to the final merged version of each relation.

138

3 IDENTIFYING THE RELATIONSHIPS

By this stage, a full relational model should have been developed. Any relational model will contain within it implicit details of the relationships between the entity types in the model (in the form of 'connector' attributes). These relationships can be extracted from the model by means of a tool called a **Key/Entity Type Matrix**.

The easiest way to illustrate the use of a key/entity type matrix is by example, but before that can be done there needs to be a more substantial relational model on which we can work. So far in this chapter we have built a small excerpt of a model, based on the attributes derived from a 'sales order' document. Figure 7.8 shows an extended version of the model, incorporating two extra entity types, 'Delivery' and 'Delivery Line'.

CUSTOMER
<u>Customer Number</u>, Customer Name, Address,
Credit Status

ORDER
<u>Order Number</u>, Customer Number, Order Date

ORDER LINE
<u>Order Number</u>, <u>Item Number</u>, Quantity, Units

ITEM
<u>Item Number</u>, Quantity-in-stock, Re-order Level

DELIVERY
<u>Delivery Number</u>, Customer Number, Delivery Date,
Route

DELIVERY LINE
<u>Delivery Number</u>, <u>Order Number</u>, <u>Item Number</u>,
Qty Delivered

Figure 7.8 The Extended Relational Model

139

3.1   The Key/Entity Type Matrix

This is a useful tool for helping the analyst identify all of the relationships which are implicitly held in the relational model. It does this by examining each entity type in turn, and using its unique identifier to search for any connector attributes in the other entity types. If a connector is found, then it indicates the presence of a one-to-many relationship between the two entity types involved. A list can then be made of all the relationships inherent in the model, and with the help of these, a model similar to that of the Data Model can be built.

The matrix is constructed by placing the names of all the entity types in the top columns: (the order in which they are placed is unimportant). The row entries consist of the names of all attributes which make up all or part of the unique key for each entity type. They can be identified simply by moving through the list of entity types in the columns, and noting down the key attribute names as they occur. Obviously the same key attribute can occur for a number of entity types; (for example, the Item Number is part of the key for three different entity types). Even so, it is only listed as one row: (see Figure 7.9).

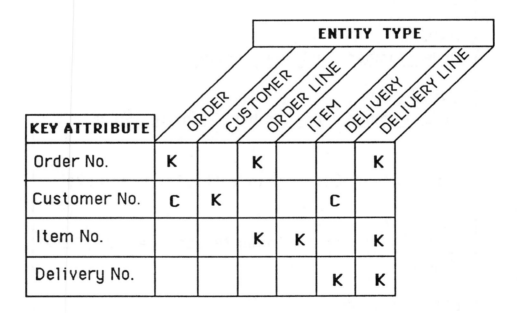

Figure 7.9 The Completed Key/Entity Type Matrix

The cells of the matrix can then be filled in:

140

1. The letter 'K' is used to represent the association of an entity type with its key attributes. 'K' is entered in the cell for which the column is the entity type and the row is one of its key attributes:(for example, the cell which intersects the 'order' column with the 'order number' row).

2. The letter 'C' is used to indicate the presence of a connector attribute in a particular entity type. These are identified by taking each entity type in turn, and checking the relational model for the presence of any of the attributes from the matrix rows.

Figure 7.9 shows how the completed matrix should look.

3.2 Extracting the Relationships

The method by which details of the relationships are extracted from the matrix is very straightforward. Each entity type (column) is taken in turn, and the 'K' cells are identified for it. The analyst then simply looks along the row(s) involved, making note of any other column in which **all** those key attributes are present (either as 'K's or 'C's). If such a column is found, its entity type is marked down as being at the 'many' end of a one-to-many relationship, with the original entity type being at the 'one' end.

Three examples should suffice to make this clear:

1. In the matrix in Figure 7.9, the first column represents the 'order' entity, and the 'K' cell indicates its key attribute 'order number'. Looking along the 'order number' row we find that the column 'order line' has a value in its cell, as does the column 'delivery line'. This indicates that the 'order' entity type has two one-to-many relationships, and they would be recorded like this:

ORDER   --------< ORDER LINE

ORDER   --------< DELIVERY LINE

2. If we take one of the later columns, the 'item' entity type, (which has 'item number' as its 'K' cell), by looking towards the end of the matrix we find a column which has a value in the same row, (the 'delivery line'), and by looking towards the start of the matrix we find another (the 'order line'). These would be recorded in a similar manner:

141

ITEM -------< DELIVERY LINE

ITEM -------< ORDER LINE

3. The third example is of a multiple-attribute key. The column
    representing the 'order line' entity type has a 'K' value in
    two rows. Examining the rest of the matrix, it can be seen
    that the only other column with values in those two rows is
    that representing the 'delivery line'. This relationship again
    needs to be recorded:

ORDER LINE -------< DELIVERY LINE

Figure 7.10 shows the full list of relationships identified from the matrix.

ORDER	————<	ORDER LINE
ORDER	————<	DELIVERY LINE
CUSTOMER	————<	ORDER
CUSTOMER	————<	DELIVERY
ORDER LINE	————<	DELIVERY LINE
ITEM	————<	ORDER LINE
ITEM	————<	DELIVERY LINE
DELIVERY	————<	DELIVERY LINE

Figure 7.10 The Relationships from the Relational Model

4 COMPARING THE MODELS

Having now identified the relationships as well as the entity types and
attributes, we are in a position to build a model similar to the intuitive data
model discussed in the previous chapter. The two models can then be

compared, and decisions can be made as to the content of the final version of the data model to be included in the Business Requirements Specification.

4.1 Building the Diagram from the Relational Model

This process again is straightforward; the most difficult part of it is deciding whereabouts on the diagram to place the boxes! For each entity type a box is drawn, and the relationships from the list are drawn in between the boxes.

The full model is shown in Figure 7.11. (It must be remembered that this has been built from attributes identified for only a small part of the system. In reality a much more thorough investigation and analysis would have had to take place.)

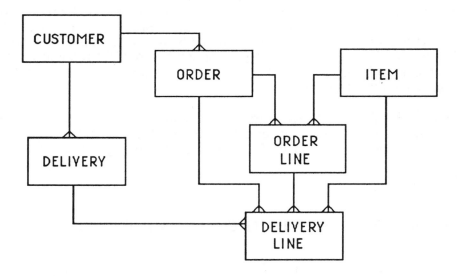

Figure 7.11 The Entity Diagram from the Relational Model

4.2 Arriving at the Final Data Model

The two diagrams, that of the intuitive data model and that constructed from the relational model, can now be compared. In theory they should be very similar, after all they refer to the same system! In practice however there are almost always differences, some of which are important and others minor. These differences must be resolved into a single data model which will be used as part of the Requirements Specification.

143

It is quite usual to find, when comparing models, that they seem to cover slightly different areas of the proposed system, and very often it is the relational version which is the larger of the two. In this case it is necessary to reach agreement with the user on the boundaries of the system, and if necessary, extend investigation of the relational model to be sure that the whole system has been checked.

The second point worth making is that in the relational version of the model there are often more relationships recorded between the same entity types. For example, the one-to-many relationships between 'order' and 'delivery line' and between 'item' and 'delivery line' shown in Figure 7.11 are unlikely to have been included in the original model.

Very often these extra relationships are important, and need to be carried into the final version. Sometimes however they do not really add anything to the model; they just make it appear more cluttered. The two relationships previously mentioned may fall into this category, in that both are already indirectly established via the 'order line' entity type. In circumstances like these the decision as to whether they should be included on the final model is left to the discretion of the analyst (though it should be recognised that those relationships are supported by the relational model whether or not they are marked on the data model).

Whatever decisions are finally made, it is imperative that the data model and the relational model should accurately reflect each other, and represent the best view of the information availably to the system being developed.

5 ADDING VOLUMES TO THE DATA MODEL

Having decided on the final version of the data model, all that remains is to record in the top right-hand corner of each of the entity types, the number of entities that on average are expected to be present in the table (see Figure 7.12). This is important information which will later be used by the designer when making decisions on the physical structure of the system's files or database. At this stage however, its purpose is simply to provide an overall appreciation of the magnitude of the information involved.

These values must be gleaned from the investigation, and must take the time factor into consideration. For example, the number of order entities should not only include those for current orders, but also those which have been satisfied and are being held for legal or statistical reasons. It may well be that the average length of the order entity table is the equivalent of three year's worth of orders!

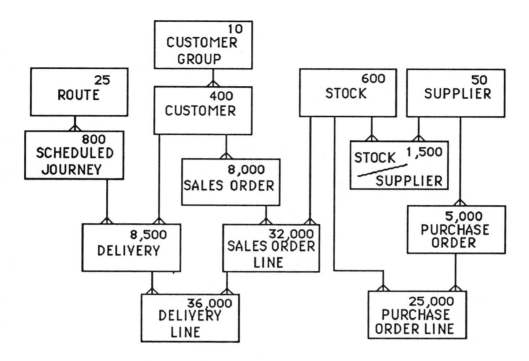

Figure 7.12 The Data Model with Volumes

6 SUMMARY

This chapter has been concerned with the process of analysing the detail of the information available to the system, in order to confirm and extend the view developed using the more intuitive approach discussed in chapter 6. This analysis involves the use of formal techniques such as Normalisation, and produces a rigorous and somewhat mechanical model, which needs to be tempered by the pragmatism and common sense of the analyst, to suit the practicalities of the situation.

Although these techniques have been discussed in some detail, it is not suggested that they will be required for all systems. What is essential is that a Data Model diagram and a Relational Model constitute part of the Business Requirements specification. None of the other supporting forms, matrices or diagrams from the data analysis process are required as part of the formal documentation of the system.

8 COMPLETING THE BUSINESS ANALYSIS

This chapter completes the description of the development of the Business Analysis stage of a project.

1. It begins by examining the relationship between Business Models for the 'existing ' and the 'required' systems (see Figure 3. 2) and illustrating how, in the latter, the separate logical views of function and data inherent in the different techniques of DFD and Data Modelling can be reconciled.

2. It goes on to discuss the main tools and techniques which would normally be used to support and document the Business Model. In this methodology, no strict standards for these are proposed, but suggestions are put forward as to the types of tool to be used, and the levels to which they should be implemented.

3. Finally, a summary of the whole Business Analysis stage is given, highlighting its relationship with the other stages and options of the systems development process.

1 INCORPORATING NEW REQUIREMENTS

1.1 Integrating the New and Existing Requirements

Normally the first versions of the business models discussed up to this point are based on a description of what currently happens in the business, and as such they represent the requirements for those aspects of the new system which are implicit in the existing system. In the case of most development projects, the bulk of the systems requirements can be obtained in this way; the user/owner usually wants the new system to provide all the functions of the old system, albeit cheaper, faster, and more accurately.

However, there are almost always some logical requirements in the form of constraints, controls, improvements or new ideas which, though they are not present in the existing system, are requested by the user or identified by the analyst during the investigation. These must be listed on some form of

147

document, and then incorporated into a new version of the business model, after the models of the existing system have been checked for accuracy.

Although the method of documenting the investigation is not considered an essential part of the methodology, an example of a suitable type of document for recording these new requirements is given in Figure 8.1, and a sample of new requirement details is shown.

NEW BUSINESS REQUIREMENTS				
Requirement	**Priority**	**Function/Entity**	**Options**	**Refs**
1. A greater control over Materials during Production	M	Distribute Production Materials	Provide usage reports to Production Mgmt.	5.1
2. A reduction in reject rates for Materials from Suppliers	H	Inspect Materials	1. Better supplier performance monitoring	3.4
			2. Reduce quality standards	

Figure 8.1 Examples of New Logical Requirements

In other circumstances, the new system may not be based on any existing system, or it may be that the logical differences between old and new are so great as to make it pointless to model the old. In these cases, only the new system is modelled, though it should be stressed that the methods used are identical to those discussed so far in this book.

On rare occasions, the user may require an entirely different logical approach to the business in the new version of the system. For example, an insurance company, which was organised to treat its different types of insurance policies as separate business functions, may decide that a better approach is to base its structure on the complete support of the customer, offering a full set of policies for each client. In such circumstances, the whole system has to be remodelled, from the Business Function Diagram onwards.

In all cases however, only one set of models represent the 'Business Model', the requirements specification to be agreed with the user and used by the designer. The new requirements are normally added into the existing model, starting from the top, the Business Function Diagram, if necessary, and adjusting the remaining models as required. These adjustments must be done very carefully, preserving the qualities of simplicity, accuracy and logic which should be present in the earlier version of the models.

Figure 8.1 illustrates a number of possible new requirements for a system. It should be noted that for each there may be several different ways of incorporating it in the system; it is the analyst's job to find the simplest and most agreeable way from the user's point of view.

In the first example, a problem has been identified concerning the control of materials between Stores and Production. In the present system the materials required for a production run are sent to the appropriate production department, and at the end of the run the unused materials are returned to stores. However, these returns are not always currently recorded, and this causes discrepancies between stock and stock records. The analyst may suggest, and the users agree, that there is a need for an extra function to monitor stock usage by the production department. Such a function can then be incorporated in the new versions of the Business Function Diagram and the Data Flow Diagram. It may also be necessary to include one or more new entity types and relationships in the Entity Model for the system, and this would obviously also lead to changes to the Relational Model.

1.2 Integrating the Function and Data Aspects of the Model

It has already been pointed out that the four main models making up the Requirements Specification each take different views of the system.

The Business Function Diagram takes a view of the functions to be carried out, ignoring the information which is to be used for the task.

The Data Flow Diagram again concentrates on the functions, but examines the information necessary to carry out the tasks involved.

The Data Model and the Relational Model both ignore the functional side of the requirements, and simply describe at different levels the information which is available to the system.

These different views of the system must somehow be brought together in order to support and cross-check each other; after all, only one system is to be designed!

It can be seen that the two function-orientated models, the BFD and the DFD, are heavily inter-dependent; in fact one is used as the basis for the design of the other, and whenever a change takes place in one the other may need to be amended to maintain consistency.

The same could equally be said of the other two models, which together give a detailed view of the data, its content and its relationships. Changes to one can again result in the need for changes to the other.

The problem lies in the inter-connection of the 'function' view and the 'data' view. The ideal tool to provide this inter-connection is the Data Flow Diagram, as it considers both aspects. However, the 'data stores' of the DFD do not necessarily correspond with the 'entity tables' of the Data Model. Many methodologies overcome this by making use of a data store / entity type matrix, which illustrates in which entity types the information for the data store is defined.

In the Systemscraft methodology a simpler approach is taken. The concepts of a **Data Store** on a logical DFD and an **Entity Type** in the Data Model are treated as the same. This means that in the final version of the Business DFD, all the data stores entries in the DFD are altered to show the appropriate entity types which are to contain the necessary data. Figure 8.2 shows a model **before** this alteration takes place, and Figure 8.3 illustrates the revised version with its Entity Type/Data Stores.

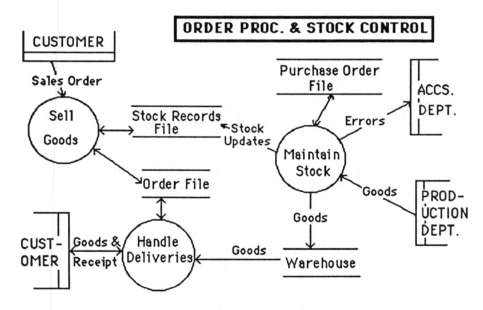

Figure 8.2 Example of DFD with Original Data Store Names

In some cases there is a simple change of name, in others there are more than one entity type involved in the replacement (some of the required

150

information is in one, some in another). Often for some data stores there is no change needed at all, because the analyst has built the DFD knowing that this integration was to take place, and has pre-empted it.

There is one minor change in the use of symbols to facilitate this integration; where two entity types would almost always be held and accessed together (as in the case of Order and Order Line), they can be expressed on the DFD as a kind of double-store. Figure 8.3 illustrates this.

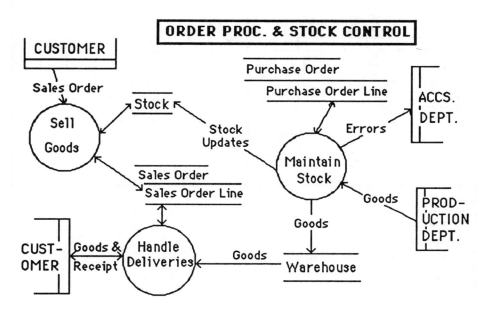

Figure 8.3 Example of DFD with Entity Types

When these adjustments have been made, it is possible to complete the check for consistency in the whole Business Model. It is now possible for example to check whether all of the entity types of the Data Model are made use of in the system. If they are not, it may mean that some entity types should be considered as being outside the system boundary. On the other hand it may also mean that some important functions or processes have been overlooked, and the function-orientated models need to be reconsidered.

Figure 8.4 gives a graphical illustration of the Business Model showing the inter-relationships of the four constituent models. It illustrates that:

1. The Functions on the BFD and the Processes on the DFD are the same,

2. The Data Stores on the DFD and the Entity Types on the Data Model are the same,

3. The Entity Types in the Data Model and the Relations in the Relational Model are the same.

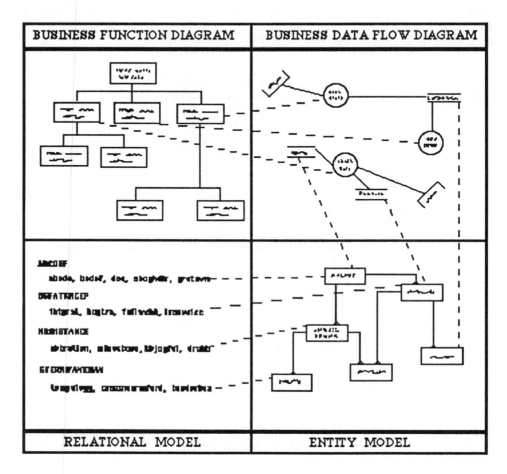

Figure 8.4 The Business Model with its Inter-relationships

2 SUPPORTING TOOLS AND DOCUMENTATION

So far, the four main components of the Business Requirements Specification have been described at some length in this book. It is however also necessary to provide a further level of supporting documentation as part of the specification. The models themselves provide the structure of the requirements, but some form of text description is needed to put the detail into this structure for both analyst and layman to examine.

Not only is support needed for the documentation of the specification; the process of developing the four models must itself be facilitated by the use of anciliary techniques. For example, it is essential that some relatively formal method of discussion and agreement is made use of throughout the whole analysis and design process.

The following group of tools and techniques should provide the necessary level of support to the specification and to the continuing process of analysis:

1. Process Descriptions

2. Data Flow Dictionary

3. Structured Walkthroughs

It will be seen later (Part 3) that these same techniques prove to be equally useful during the Systems Design phase of a development project.

2.1   Process Descriptions

The logical processes at the bottom level of the Business Function Diagram and the Data Flow Diagram may actually represent a quite complex set of business requirements. The models themselves can only convey an overall impression of these requirements; there is clearly a need for a more detailed description of each of these processes.

While this description should be as **informative** as possible, it must also be **concise, accurate, unambiguous** and **unemotional**. It should indicate a series of logical activities, what they consist of, and the order in which they are obeyed.

There are a number of different tools made use of for this purpose in various methodologies (for example, process flowcharts, decision tables, action diagrams, etc.), and each has its strengths and weaknesses. The particular tool recommended in this methodology is one of the oldest, simplest and most proven. It is known as 'Structured English'.

PROCESS DESCRIPTION   [Logical|~~Physical~~]

Process Name   Take Order   **Ref.**   1.2

Accept Order-Details from Customer
Check validity of Customer against Customer file
For each Order-Line
    Check validity of Item-Number
    Check availability of stock against Stock file
    Calculate cost for Order-Line
    Add Order-Line cost into Order-Cost
Check Order-Cost against Customer-Credit-Limit

Figure 8.5 Example of a Logical Process Description

PROCESS DESCRIPTION   [~~Logical~~|Physical]

Process Name   Take Order   **Ref.**   1.2

Allocate next available Order-Number to Order
Accept Order-Header-Inf. from Customer
Check validity of Customer against Customer file
If Customer Invalid
    then output error message 5.6
            request re-submission of Order

For each Order-Line
    Accept Order-Line from operator
    Check validity of Item-Number against Item file
    If Item-Number invalid
        then request re-submission of Item-Numb

Figure 8.6 A Process Description for a Physical Process
(from the Design Stage)

154

Structured English has been described as a true 'pidgin' language, in that it is made up of the vocabulary of one language (English), and the syntax and structure of another (structured programming language).

The English used is a small and deliberately simplified subset, which consists of the following:

1. English verbs which lend themselves to the imperative mode (eg. 'Do This', 'Create That', etc.)

2. Terms defined in the Data Flow Dictionary, the relational model, or other models in the specification,

3. Certain reserved logic words (used to show the syntax and structure).

It excludes all qualifiers (ie. adverbs or adjectives), compound sentences, and emotive expressions.

The activities of the business processes must be described by using one of three constructs which were originally developed for use in structured programming. These are:

SEQUENCE
SELECTION
ITERATION

1. The SEQUENCE construct simply means that the statements describing the process occur one after the other, and it is assumed that such statements are obeyed in order.

2. The SELECTION construct is used whenever a decision is made in a process, and two or more options have to be described. A basic statement structure of the format IF.....THEN....ELSE is used.

3. The ITERATION construct is used whenever a particular series of activities is to be repeated a number of times within the process. Again, there is a formal structure to the statement group. It starts with the phrase FOR EACH, followed by a circumstance, then by a number of indented statements. The non-repeating statement following the group should not be indented.

Examples of these constructs occur in Figures 8.5 and 8.6, which represent process descriptions from different stages of the development of a project.

Although the process description is expressed in this formal language, every attempt should be made to make it as readable as possible. There is a certain amount of flexibility in the use of the tool, in that the original process description may be relatively woolly and general, but over time, as the specification develops and requirements are clarified, it may gradually be improved.

Finally, it is important to realise that this tool can be used not only for describing the LOGICAL processes of the business analysis, it is also used to describe the PHYSICAL processes of the systems design stage. A process description will in fact form the main part of a specification for each proposed software component, whether it be module, routine, or full program. Figure 8.6 illustrates this. It represents a specification for the Order Acceptance computer program, and it contrasts with Figure 8.5, which represents the logical requirements specification for the same function. One obvious difference between the logical and physical descriptions occurs in the data stores. In the logical version these are known as 'Tables' and 'Entries', whereas in the physical process description they become 'Files' and 'Records'.

2.2 Data Flow Dictionary

The general topic of data dictionaries (already discussed briefly in chapter 1) is becoming more and more important in the field of systems development. Modern forms of data dictionary are supported by substantial software packages, and provide many facilities, ranging from the simple recording of systems names through to the automatic validation of data being put into the system. The most advanced, often referred to as 'active' data dictionaries, are able to generate database 'schemas', and even actual program coding from the detailed analysis and design specifications entered by the analysts.

These data dictionaries differ in the standards used for definitions, and are dependent to a large extent on the particular hardware/software environment of the user. In terms of the discussion of this methodology they are outside our scope; users of the methodology will adapt their approach to documentation (and to program generation) to suit the equipment and packages available.

It is however necessary in any methodology to provide a simple explanation of how the more detailed information of the system might be recorded in the formal requirements specification. Much of this has already been explained; the relational model obviously provides the detail of the entity tables and their content, and the process descriptions contain the level of information necessary to describe their purpose. The one type of term for which so far there has been no opportunity for descriptive definition is the DATA FLOW from the DFD, and it is the purpose of this section to outline a method for doing so.

All the data flow names mentioned anywhere in the logical data flow diagram should be described in some detail, and included in alphabetical order in what is referred to as a 'data flow dictionary'. These entries in the dictionary will describe the contents of the flow in terms of the smaller data items (or fields) of which they consist. Figure 8.7 gives an excerpt from such a dictionary.

```
┌─────────────────────────────────────────────────────────┐
│          DATA FLOW DICTIONARY                           │
├─────────────────────────────────────────────────────────┤
│                                                         │
│  Customer-Detail  =  Customer-Number +                  │
│                      Customer-Name   +                  │
│                      Customer-Address                   │
│                                                         │
│  Order-Header     =  Order-Number    +                  │
│                      Order-Date      +                  │
│                      Customer-Detail                    │
│                                                         │
│  Order-Line       =  Item-Number     +                  │
│                      Item-Description +                 │
│                      Item-Quantity-Ordered              │
│                                                         │
│  Sales-Order      =  Order-Header    +                  │
│                      { Order-Line }²⁵₁                  │
```

$$\{ \text{Order-Line} \}_1^{25}$$

Figure 8.7 Excerpt from a Data Flow Dictionary

The symbols and conventions used here are those put forward originally by DeMarco in his book *Structured Analysis and Systems Specification*. They are generally considered to be the simplest, and are used extensively by major organisations.

It should be remembered that these definitions are of 'logical' data flows, and therefore do not represent specific documents (either from the old or the new systems). Having said that, it is not uncommon for an analyst to use a

157

document (for example a sales order form) to help in the identification of the components of a flow.

It should be noted that there may be more than one dictionary entry per flow, in that some flows are quite complex, and some of the elementary data fields may also need to be grouped under broader definitions. Figure 8.7 illustrates this point, in that all the entries shown represent components of one particular data flow, 'Sales Order'.

2.3   Structured Walkthroughs

A structured walkthrough is a mixture of a meeting and a presentation, but it is entirely different from any other form of either of these, as used during the systems development process. In a structured walkthrough, a model or a part of a model is put forward for discussion and agreement to a team of interested parties. The model for discussion should have been already completed to the satisfaction of the analyst who is building it, and the purpose of the walkthrough is to get the whole team to look for weaknesses or potential improvements to it before accepting group responsibility for it as a satisfactory piece of work.

The approach has been developed from a technique originally put forward by Gerald Weinberg in his book *The Psychology of Computer Programming*. That technique was known as 'Egoless Programming', and consisted of programmers putting forward their work for critical examination by their peers. The finally accepted version of the work was considered not to belong to the individual programmer, but to the group.

The secret of a successful walkthrough is in the attitude of the people present: the comments and criticism during the walkthrough must not directed at the individual whose piece of work is being discussed, but at the work itself.

Also, the members of the group should not attempt to re-design the model during the walkthrough; their job is simply to point out any weaknesses and put forward suggestions for consideration by the analyst who is submitting the work. It should not degenerate into a 'brainstorming' session! If there are any serious flaws, the group should ask the analyst to re-submit after changes have been made.

However, when the model is found to be acceptable, the group will 'sign it off'. Eventually the whole of the Requirements Specification should be signed off in this way, (see Figure 8.8).

It will be seen later in the book that the structured walkthrough is an absolutely essential aid in the management of any EDM project.

Figure 8.8 Walkthroughs for a Requirements Specification

For any walkthrough the members of the group who are to conduct it are required to adopt formal roles:

1. The CO-ORDINATOR should make sure that all the papers are circulated before the meeting, and is responsible for making all the necessary arrangements in order that the meeting can be carried out. The Co-ordinator should start the meeting by stating its objectives, and, while it is in progress, should control it by monitoring the time, encouraging contributions from everyone, and making sure that all

159

problems and decisions are recorded. The co-ordinator should NOT get involved in the detail of the discussion (eg. defending the model) but, at the end of the meeting, should arrange the 'signing off' of the model, if it is to be accepted.

2. The SECRETARY should take notes throughout, and when required should summarise the points made. Most importantly, any final decisions about further action to be taken must be recorded. Again, the secretary should not be a contributor to the discussion, other than to ask for clarification of points made.

3. The PRESENTER should give a simple un-biased description of the model (for example, by tracing the flows through a DFD), providing maximum opportunity for discussion and comment. The presenter must avoid the temptation to defend the model as though it was a personal responsibility, but should provide clarification when necessary, and describe as required the reasons behind some of the decisions.

4. All the rest of the team are REVIEWERS. They are all considered as equals: there should be no managers present (at least not in their role of making staff judgements). Some of the reviewers may however be user representatives, but they will be present only to help clarify requirements. No attempt should be made to 'sell' the system at a walkthrough; (there are other types of meeting and presentation for that purpose). All reviewers should make some contribution, whether it be a request for clarification, a suggestion, an insight or an affirmation.

The standard text on this subject is Edward Yourdon's *Structured Walkthroughs*, and readers who are interested in exploring the topic further are advised to obtain that book.

3   SUMMARY OF THE BUSINESS ANALYSIS PROCESS

This book has described in some detail the tools and techniques used in the Systemscraft methodology to complete the analysis of a business system, and to provide a full Business Requirements Specification for the users to consider and accept. The specification is also the main source of information for the designers of the new system; not only does it explain to them their objectives, but it provides the scope and constraints within which they must work.

In this summary three aspects of the business analysis method are considered:

1. The use of the methodology to describe itself,

2. The format of the Requirements Specification,

3. The interface with the systems design stage.

3.1 The Methodology Describing Itself

It has often been pointed out that the use of the term 'methodology' to describe the systems development procedure is incorrect: methodology is 'the study of methods'. However, one of the arguments for using the term is that the bundle of tools held together by a framework is in fact a method for describing systems. A system is itself a method of organising and carrying out an activity, so the method of describing it is a 'method for describing methods'.

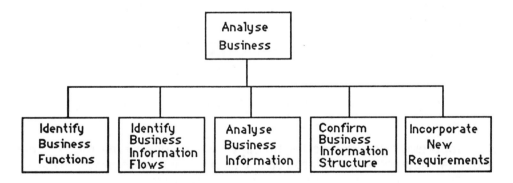

Figure 8.9 A Business Function Diagram of the
Business Analysis Stage

Another way of expressing this is that the systems development methodology is itself a system. Well, as it is also a 'method of describing systems' surely it must be possible to use it to describe itself!

It is in fact common practice for one of more of the methodology tools to be used for this purpose. Usually however , in order to be of any practical use, the description needs to be physical rather than logical, so those tools which

161

provide physical modelling capabilities are considered to be more suitable to the task. Figures 8.9 and 8.10 illustrate the Business Analysis stage of the Systemscraft methodology in such a form.

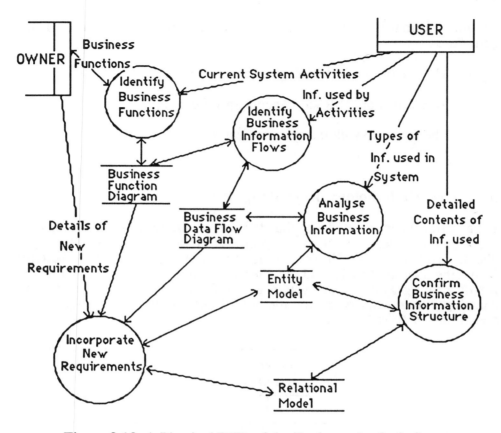

Figure 8.10 A Physical DFD of the Business Analysis Stage

One important potential use for this approach of describing the systems development process is that it can enable a senior analyst or project manager to describe any adjustments that he or she wishes to make to the methodology to tailor it for a particular project. (For example, if there was no existing system on which to base the analysis, or if there was a requirement to make minor adjustments to an existing computer system, then the DFD model in Figure 6.18 would not give a true reflection of the required approach.)

This facility is an important part of the flexibility and scalability aspect of the methodology. It means that it can be tailored for each different systems development project to be taken on. It is of course more important in the planning of the systems design approach, where there can be a number of widely differing options, whereas, by comparison, the business analysis approach is relatively prescriptive.

3.2 Format of the Requirements Specification

At the end of the Business Analysis stage a formal requirements specification document should be produced. In this methodology, it should take the form suggested in Figure 8.11, and should be officially agreed by the user/owner of the system.

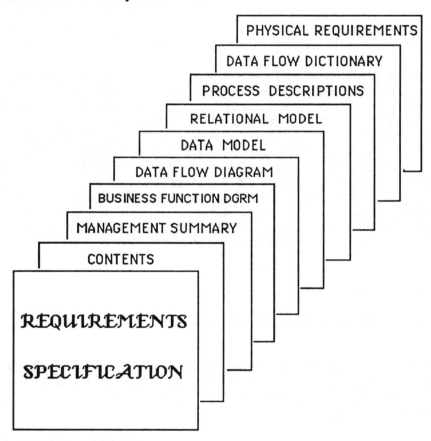

Figure 8.11 The Component Parts of the Requirements Specification

It should be remembered that by the time the specification is complete, prototyping techniques may have been applied to many parts of the system, with the effect that the project may be in the advanced stages of design. Individual analysts will have been allocated components of the system to work on, and there is a danger that as this work goes ahead, some aspects of the interface between these components may be overlooked.

163

A **Co-ordinator** must be appointed, with responsibility for overseeing the production of the full business model and the requirements specification.Whenever a change is made to part of the system, the co-ordinator must be able to assess the implications for the rest of the system, and take action accordingly. One of the most important ways of exercising this kind of control over the systems development is through the use of formal walkthroughs, (see section 2.3).

The Requirements Specification content and structure is illustrated in Figure 8.11. The format of all the component models has already been described in this book, though for the purposes of the specification report it may be necessary to annotate some of the models with a level of simplifying comment. Only the 'management summary' needs further clarification.

A flexible approach is taken to the content of the management summary; no formal structure is suggested. Everything should depend on the nature of relationship with the user/owner, and on the judgement of the analyst. If for example the user has shown very little interest in the technical aspects of the system, then the summary should comprise the major component of the specification, and the models should be reduced to the role of appendices. The summary should attempt to precis and simplify the main findings from the model in layman's terms. On the other hand, if the user has fully participated in the analysis, has signed off some of the models after walkthroughs, and has been involved in the consideration of prototype designs, then the management summary can be a brief review, with heavy references to the diagrams elsewhere in the report.

3.3   Interface with Systems Design

The production of the requirements specification represents the end of the business analysis stage. It provides the users with a definition of the business needs as seen by the analyst, and it enables them to check its accuracy and completeness.

However, the requirements specification by itself is INSUFFICIENT to place before the user/owner. Almost certainly the people who have paid for the study to be done will expect to see suggestions and proposals for the computerisation of the system. They will want to be able to choose between different options, they may even want to consider the possibility of a package solution. They may also be keen to inspect prototypes of some of these alternatives.

It is one of the basic precepts of the evolutionary development approach that such material will be available at this time, and in fact many of the design decisions will have already been made by the user.

Part 3

9 THE INITIAL DESIGN PROCESSES

Whereas the Business Analysis stage consists of a small number of modelling techniques, all of which are essential to the structure of the Requirements Specification, the Systems Design stage, though apparently more complex, and containing more different types of technique, is actually less prescriptive. There is much more flexibility as to which techniques to use, and where (and whether) to use them. Another aspect of the flexibility of the approach is that several of the models can be developed to different levels of complexity, depending on the nature of the particular project.

Because of this, the techniques used in the Systems Design stage of the Systemscraft methodology are presented in less detail than those of the earlier stage. It is assumed that some analysts wishing to take an evolutionary approach may be more familiar with other techniques, and they are encouraged to consider whether those techniques could be substituted for the ones shown here. This concept of 'tailoring' the methodology to the specific preferences of the individual IT department is an important part of the evolutionary philosophy.

An overview of the complete Systems Design stage has already been provided in chapter 3. This chapter examines in some detail the first two steps of the stage;

1. Identifying the Computer System; The use of the Systems Data Flow Diagram

2. Designing the Human/Computer Interface for the proposed system, and building the Initial Prototypes.

It should be remembered that one of the most important aspects of any evolutionary methodology is that the design process can start while the analysis is still taking place. In fact it is recommended that it should be started as early as possible in order to take full advantage of opportunities to build prototypes. **The two steps which are discussed in this chapter are exactly the ones which overlap almost completely with the analysis stage!**

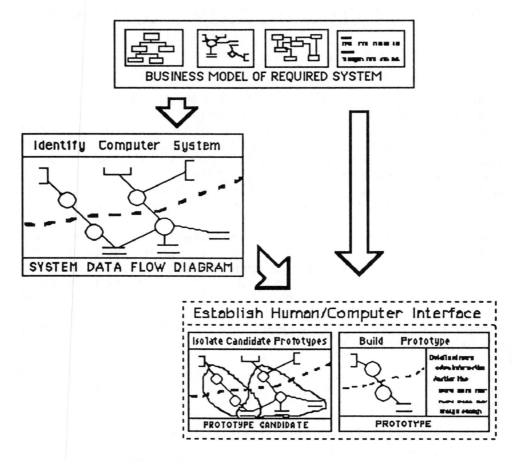

Figure 9.1 Initial Steps of Design Stage

The design process uses as its main input the Specification of Requirements constructed during the analysis process, and consisting of

> Business Function Diagram
> Business Data Flow Diagram
> Entity Model
> Relational Model
> Supporting Documentation, including:
> Process Specifications
> Physical Requirements Chart
> Data Dictionary.

As well as this, there needs to be a full and continuous dialogue conducted with user and owner throughout the design process. This may take the form of further fact-finding interviews, walkthroughs and presentations. One of

170

the most important ways in which this dialogue can take place is via a prototype, and the methodology is geared to providing prototyping opportunities at a number of stages in the design process. These will be highlighted as the description progresses.

Figure 9.1 (which is an excerpt from Figure 3.3) illustrates these first two processes involved in the systems design. Each will now be described in more detail.

1   IDENTIFYING THE COMPUTER SYSTEM

This is the first stage in the systems design process, and its purpose is to identify which parts of the proposed system are to be handled by the computer and which are to carried out by the user. The method involves using the Business DFD from the specification of requirements, and working through all the 'bottom level' processes on it, considering what role the computer should take in each of them. The results of these considerations are recorded on a major new model, known as the 'Systems Data Flow Diagram'.

Very commonly, the designer may come up with a number of different ways in which the computer could be used in a particular series of processes. These will then be discussed in detail with the user, and agreement will be obtained as to the preferred option. The agreed computer boundary can then be formalised on the Systems DFD, new process descriptions can be created, and other models can be modified to bring them into line. Figure 9.2 gives an example of a Systems DFD. It describes one of the main purchasing functions in a large organisation, that of developing a purchasing agreement with a supplier.

1.1   The Systems Data Flow Diagram

The Systems DFD tool is almost identical to that used for the Business DFD during the analysis stage; this is in keeping with the general policy of using the minimum set of different types of modelling tool within the methodology. Also, it is imperative that the users should be able to understand completely the ideas recorded in the Systems DFD, so there are major advantages in using a set of symbols with which they are already familiar.

Whereas the Business DFD is essentially a 'logical' model of the required system, the new Systems DFD can be better described as 'semi-physical', because although it states which processes are to be performed on the computer, it does not concern itself with how this is to be done, which programs and computer files are to be created and made use of, etc. (This occurs later in the methodology, and a much more physical model, known as a 'Computer Data Flow Diagram' is used).

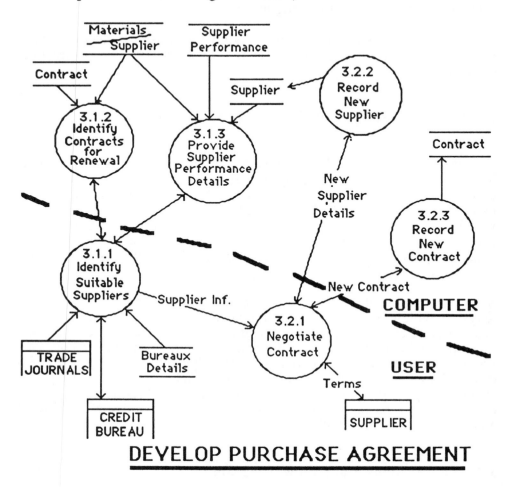

Figure 9.2 Example of a Systems Data Flow Diagram

1.1.1 The Symbols Used

As can be seen from the example in Figure 9.2, the standard DFD symbols are used to represent processes, data flows, data stores, and internal or external agents. The main difference is the use of a dashed line through the diagram separating the computer and user processes. The diagram is also marked to show which side of the line is which. The dashed line is known as the 'Computer Boundary'.

172

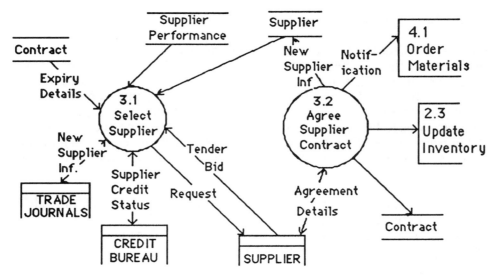

DEVELOP PURCHASE AGREEMENT

Figure 9.3 The Business DFD for Figure 9.2

1.1.2 The Splitting of Logical Processes

The processes in the new Systems DFD are derived directly from the logical processes in the related part of the Business DFD. In fact, if one of the Business processes is carried forward into the design as either a completely computerised task, or as a completely non-computerised task, then the process name will be the same on both the Business and Systems DFD models.

On the other hand, it is often the case that a logical process will be designed to be carried out partly by computer and partly by the user. For example, Figure 9.3, which shows a Business DFD for the development of purchasing agreements with suppliers, is the source from which the Figure 9.2 Systems DFD was designed. Here it can be seen that the logical function, 'Select Supplier' has been converted into three processes, one to be done by the user ('Identify Suitable Suppliers'), and the other two by the computer ('Identify Contracts for Renewal' and 'Provide Supplier Performance Details'). Similarly, the logical process 'Agree Contract' has become three physical processes, one to negotiate the new contract and two to record details of agreements on the computer.

It is imperative, however, that any new process on the Systems DFD can be traced back to the Business process from which it has sprung.

1.1.3 Numbering the Processes

One of the most effective ways to establish the connection between processes in the Business DFD and the Systems DFD is to relate them by process number. The standard approach to this is to give any process which is carried over unchanged (and therefore with the same name) the same number. For example, if it had been decided that the process 'Agree Contract' was to be done without use of the computer (or, more unlikely, if it had been decided that it was to be done completely by computer), then the process could be carried over into the Systems DFD with its existing name and number.

When any business process is split to go into the System DFD, then all the new processes should be given the business process number plus a subscript. Figure 9.4 shows how a numbered business process would be split and renumbered on the Systems DFD.

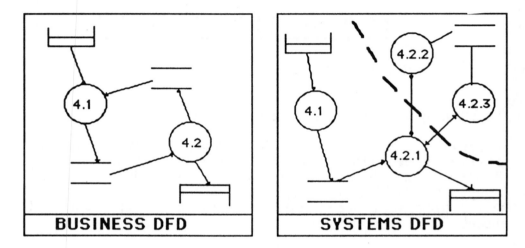

Figure 9.4 Numbered Processes on Business and Systems DFDs

This numbering approach emphasises the inter-dependence of the processes on the two models, by suggesting that the split processes are one level down in the hierarchy from the process being split. In one sense they can be thought of as an extra level attached to the bottom of the Business Function Diagram model, but it must be stressed that they can not be added to the formal BFD model! They are not functions but processes!

The use of subscripts for split processes can occasionally lead to the apparent anomaly of different levels of subscript occurring on the same DFD page. This is not considered to be a problem in a Systems DFD.

1.1.4 Data Stores

The data stores on the 'computer' side of the computer boundary obviously must represent entity types from the data model completed as part of the specification.

On the other hand, the data stores from the user side of the boundary represent clerical and other files, not to be held on the computer. Any information which is to be held manually rather than on the computer should be taken off the design version of the Data Model. This means that some of the entity types may disappear from the model, and others may have their attribute list reduced. This activity clearly needs to take place before major decisions are made about the physical file/database design.

It should also be noticed that some of the 'internal agent' references may not appear on the new diagram. This is because information on the computer side of the boundary is more likely to be put into mutually accessible files than passed from process to process by means of a physical data flow.

1.2 Building the Model

The building of the Systems Data Flow Diagram represents the recording of the systems designer's intentions concerning the use of the computer within the proposed system. The DFD tool itself is an analysis and design aid, enabling the designer to model ideas quickly, and perhaps discard them after more careful thought. This continuous process of mooting and modifying, considering and rejecting, and then providing options for discussion, can best be described using a detailed example.

Figure 9.5 shows a Business Function Diagram representing the 'Purchasing' function of a manufacturing organisation. The design of part of this system, the development of purchasing agreements, has been considered earlier in the chapter. We are now going to examine how the designer might put forward and record options for computer involvement in the rest of the system, the creation and monitoring of purchase orders.

The Business DFD for the process 'Handle Purchase Orders' is shown in Figure 9.6. It consists of two sub-processes, one of which involves the creation of the purchase order, and the other the tracking of the order's progress when problems or delays occur.

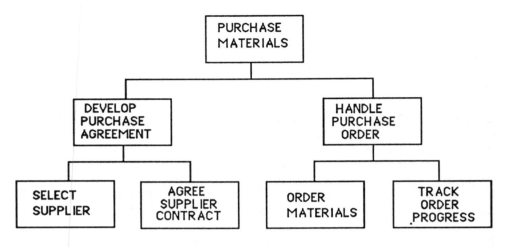

Figure 9.5 Business Function Diagram for 'Purchase Materials'

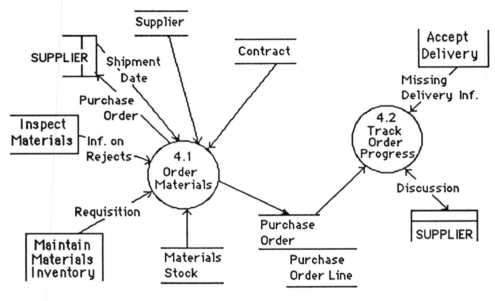

Figure 9.6 Business DFD for 'Handle Purchase Orders'

The construction of the order takes place when there are a number of material stock lines below the re-order level, and several of them can be provided by the same supplier. An order is created and sent to the supplier, who then provides details of expected delivery dates. (Sometimes the delivery dates are agreed with the supplier before the order is made.)

If orders are not delivered by the agreed date, or if a sudden shortage of materials is experienced (perhaps as a result of un-anticipated customer demand), discussions take place with the supplier to expedite the delivery. This can involve adjusting order quantities, agreeing to part shipments, and even to the placing of another order with a different supplier. Also, any delays in deliveries, or other form of adverse behaviour by the supplier, should be noted down.

The designer might begin by examining the 'Order Materials' process, and considering how the computer can be used to assist. First of all, the event that 'triggers' the process (ie. causes it to take place) would be identified. Normally an order will not be placed unless there is a shortage of a number of line-items which can be provided by the same supplier, though every now and then there may be an emergency one-line order.

In traditional purchase order systems, when a line falls below its re-order level, a requisition document is produced and passed to the materials buyer. Requisitions are held in the Purchasing department until there are sufficient to make up a purchase order. However, 'requisitions' are physical rather than logical, used simply to pass information within the organisation; logically the same information could be obtained by examining the materials entity table. For this reason, no requisition data store appears on the Business DFD.

There are two obvious ways in which the order creation could be set in motion, and the designer will give consideration to both.

1. There could be a **batch** program run perhaps once per day at a fixed time. This would examine the materials file for re-order candidates, and automatically build orders to deal with them all.

2. A more **interactive** approach could be taken, whereby the user entered into a dialogue with an on-line program. This could be run at any time, and would give the user the extra flexibility to make decisions on which items should go on an order (perhaps overriding default options taken by the computer program).

1.2.1 First Alternative Design

Figure 9.7 shows an example of a Systems DFD where the first of these two options has been taken. Notice the absence of dialogue between computer and user processes; the standard computer department procedure to run the batch program once per day is not shown on such a diagram. An output from the 'Create Order' process sets in motion the clerical dispatch of the order document to the supplier, and the only input to the process is the delivery date information, provided by the supplier several days after the order document was produced (and therefore put into the computer during a later run of the program). This is likely to be in the form of a sorted file of order updates.

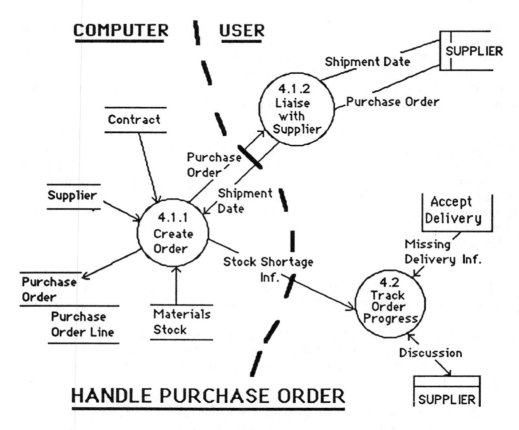

Figure 9.7 Draft for Batch-Oriented Systems DFD

The proposed design suggests that the 'Track Order Progress' process should be handled clerically, and should be triggered by the receipt of a list of critical stock shortages caused by over-due deliveries. This list would be produced daily by the computer as part of the 'Create Order' process output.

Note that the Systems DFD model of the design does not carry all the information needed for describing the option in detail to the user. It must be supported by Computer Process Descriptions, written in structured English, and data dictionary entries for each of the flows. It is also wise to provide process descriptions of a similar type for the clerical processes identified in the model.

1.2.2 Second Alternative Design

A designer who wished to consider a user-interactive computer system for building purchase orders, might come up with a solution similar to the one shown in Figure 9.8. Here the emphasis is very much on the user's contribution. In fact, in this option, the process called 'Initiate Purchase Order' is a user process, with the computer process 'Set up Purchase Order' providing support. A separate computer process simply records the shipment dates, agreed between the buyer and supplier in the 'Liaise on Shipment Date' user process.

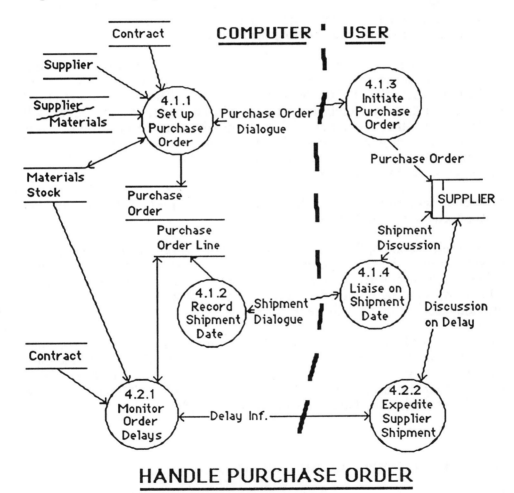

Figure 9.8   Systems DFD for the On-line Alternative

In the 'Initiate Purchase Order' process, the user will begin by asking the computer for a list of line-items which require to be re-ordered, and then request details of possible suppliers for each of the line-items on the list; (this can be obtained from the Supplier/Materials entity table).

Having browsed through this information, the user (who will almost certainly be the materials buyer) will select one supplier who can provide a number of the required line-items, and by means of a dialogue with the computer process will build a purchase order.

This approach has the advantage of giving the buyer the flexibility to choose particular suppliers in particular circumstances, for instance when an urgent delivery is required at short notice. Similarly, when a special type of material must be available by a certain date, the order for that material can be split between two different suppliers, thereby increasing the likelihood of at least one delivery being made before the critical date.

The second pair of systems processes derived from the original 'Order Materials' business process deal with the supplier's response to the receipt of the purchase order, and the buyer's reaction to it. The supplier is required to suggest a date by which the delivery will be made, and the buyer will enter this into the computer system. It may be that the proposed date is unsatisfactory, in which case the buyer may have to consider a number of options, including the cancellation of the order. Alternatively, part-delivery by an earlier date may be acceptable.

As can be seen in Figure 9.8, the 'Track Order Progress' function can also be supported by an on-line interaction.

Clearly this second option provides the user with a completely different type of system than the first, but considerations of cost and convenience will influence the user's decision. It must be emphasised that the decision about which option to take is the user's not the analyst's; the analyst is there to provide information and advice.

Both options should be put to the user by means of walkthroughs rather than other forms of presentation, and the options should be put in such a way as to allow the user to 'mix and match' components. For example, it may be that the user accepts the on-line approach to 'Create Purchase Order', but prefers the batch version of 'Track Order Progress'.

1.3   Further Design Considerations

It has been mentioned already that the purpose of this part of the book is to describe in detail the use of modelling techniques for systems analysis and design. Because of this, the discussion in this chapter has centred on how different options are recorded on a DFD rather than on how the designer may formulate ideas, what criteria may be used in deciding which options to put forward, etc. The skills needed for the latter are taught on various types of training course, and are developed gradually through experience. However, there are a few points relating to design consideration which are worth making at this stage of the methodology description.

1.3.1 Batch or On-Line

Firstly, let us consider how the designer might approach the problem of whether to put forward a batch or an on-line option, or both. Whereas only a few years ago a batch approach to design was considered to be the norm, now the designer is more likely to think in terms of an on-line system, and batch solutions are becoming rarer.

Sometimes the batch approach is only suggested when user constraints force the analyst's hand; for example when the new system must fit in with existing hardware and software, and the new system has lower priority than other current systems.

There are however three types of situation where a batch process is likely to be the better choice:

1. Where a number of transactions need to be assembled in order that a process can be carried out;

    eg. vehicle or delivery scheduling; (all deliveries need to be examined before the best routes can be decided on).

    eg. the purchase order system just examined (though as we have seen, an on-line process may provide greater flexibility.

2. Where the process must occur at fixed times;

    eg. payroll systems involve weekly and monthly payment runs; there is no real benefit in having an on-line system for making staff payments (though it may be worth considering an on-line update for amendments such as staff appointments, overtime notifications, etc.).

3. Where the process involves examining every record in a file in order to check whether there needs to be some action taken;

eg. a reminders process, or a process for issuing statements, where a date or some other attribute can trigger the production of some document.

1.3.2 Designing for Packages

When the solution proposed for the system is a package, there are extra constraints to bear in mind. The designers must have a full knowledge of the package's capabilities, and how flexible it is. They must then model the package facilities as computer processes on the Systems DFD, and illustrate how the package carries out the systems requirements as shown on the Business DFD. The processes where the user interacts with the package will also be shown on the model, as well as the flows representing the dialogue.

Sometimes the designer may have to consider a number of alternative packages for the system. The Systems DFD is an ideal modelling tool for this task, enabling each of the packages to be fitted onto the specified requirements. This should give a clear picture of advantages and disadvantages of each, and should assist in explaining the differences to the user.

Often a package must be modified before it is able to satisfy the full set of user requirements. Sometimes this modification is done by the package suppliers, and sometimes the alterations are done in-house. Whichever of these alternatives is taken, the Systems DFD and its supporting documentation will provide the basis of the physical requirements specification for the alterations.

1.3.3 End-User Design

The use of fourth generation languages and high-level query languages in the systems development process can enable some computer-articulate users to create small programs of their own. Such users may choose to build for themselves a set of programs to handle a number of enquiries or minor updates. When the Systems DFD is being developed, opportunities for this kind of user involvement can easily be identified.

2 ESTABLISHING THE HUMAN/COMPUTER INTERFACE

This is a very important step in the evolutionary process. It initiates the physical design of the system, and it does this by:

1. Isolating Candidate Prototypes; this incorporates the project management task of separating the development of the computer system into components which can be worked on independently by small analyst/user teams.

2. Implementing Initial Prototypes; these actually represent the first of a number of versions of each prototype, and this initial version will concentrate on the human-computer dialogue and the very basic functionality of the system component being prototyped.

2.1 Identifying Candidate Prototypes

The purpose of and justification for a prototyping approach has already been covered in some detail in this book. The decision whether or not to use a prototyping approach for a particular project will have been taken some time during the feasibility study, and that decision will be based on various factors, including the availability of a suitable hardware/software environment, on the attitude and commitment of the users, on the overall nature of the system to be developed, etc. Nevertheless, there is a great deal of flexibility in the implementation of the approach, and the designer will be required to make judgements as to where and when to apply the technique as soon as the computer system boundaries have been agreed.

As soon as part of the Systems DFD has been built from its Business DFD counterpart, the designer is able to examine it with a view to deciding which parts of it can be prototyped, and of what each prototype should comprise. The decision maker may need to have access to other information than the Systems DFD, in order to be able to judge the level of complexity of a potential prototype. This information may include the process specifications of the computer processes, details of any of the relevent 'physical requirements' provided with the systems specification, and perhaps some examples of the documents used in the existing version of the system. It may also be worthwhile to talk again to the user of this particular part of the system.

A candidate prototype should consist of a combination of at least one process on either side of the computer boundary, the data flow(s) which connect them, and any files accessed by the computer process(es). Normally the main data flow between the computer and user process is a dialogue; (it is rarely worth prototyping anything other than an on-line process).

In general the designer should try to keep prototype candidates small; usually they can be restricted to one user process which may have a dialogue with one or two computer processes. A prototype should be simple enough for the designer to build quickly, but complex enough to engage the user, and provide worthwhile feedback for the design team. The best balance of simplicity and complexity will depend on circumstances like the attitude, capability and availability of the user, the experience of the analyst, whether a rapid or evolutionary prototype is being used, and the criticality of the part of the system being considered.

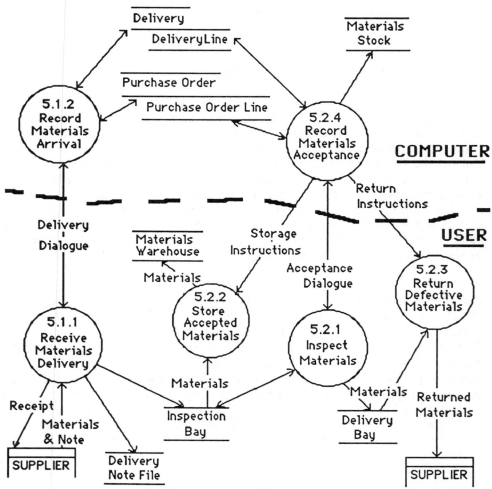

ACCEPT MATERIALS DELIVERY

Figure 9.9 A Systems DFD for 'Accept Materials Delivery'

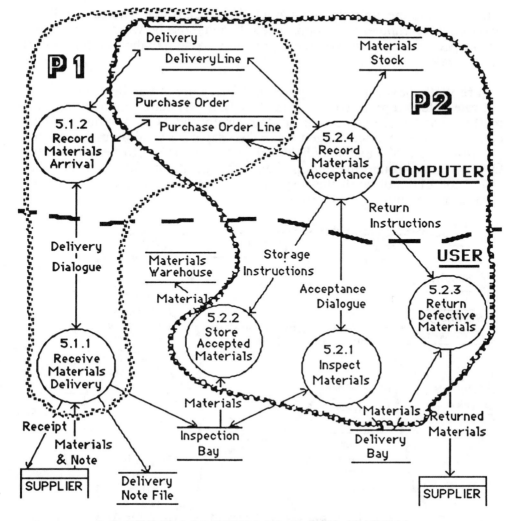

ACCEPT MATERIALS DELIVERY

Figure 9.10 Candidate Prototypes for 'Accept Materials Delivery'

Figure 9.9 shows a Systems DFD for part of a materials processing system. This deals with the acceptance of deliveries of materials from suppliers, the inspection of the materials for quality, and the return of unsatisfactory materials to the suppliers. The Systems DFD represents a proposed on-line approach, where the user in charge of 'goods in' will access the computer to check against the purchase order and record the delivery. Later on, after the materials have been accepted, the inspector records their acceptance in the computer system, which notifies staff in the warehouse (by means of a remote printer) to either store them away or return them to the appropriate supplier. This means that whereas the flows between the computer and 'goods in' clerk, and those between the computer and inspector, are interactive, the information passing to the warehouse staff is one-way.

Figure 9.10 shows the same Systems DFD, only it has been marked to indicate potential prototype opportunities. In the Systemscraft approach the method used to indicate a candidate prototype is to surround all its components by a thick jagged line. Two separate candidates have been identified, both of which may be developed, perhaps by different designers. One is centred round the dialogue with the 'goods in' clerk and the other round the dialogue with the inspector. Notice that the main files of the system are common to both prototypes.

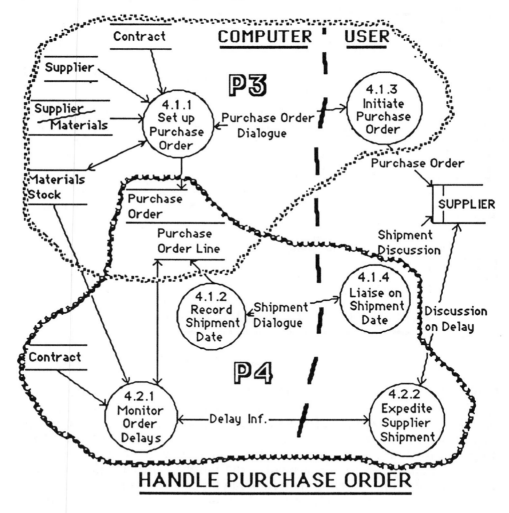

Figure 9.11 Candidate Prototypes for 'Handle Purchase Order'

Strictly speaking, the production of output to the two warehouse processes need not be included in this second prototype. However, there are advantages in producing report layouts for these, and checking them with the warehouse staff who will use them. If this approach is taken, then the jagged line would include the warehouse processes as shown.

This marking of prototype boundaries on a copy of the Systems DFD enables the senior designer to plan and schedule an approach to the development of the whole system. When the overall system is complex, and there are a number of designers involved in the project, it helps to avoid overlap of duties, and ensure that all parts of the system are covered.

A second example of the process of isolating candidate prototypes is provided in Figure 9.11. This is based on one of the Systems DFDs developed earlier in the chapter, and it shows the handling of purchase orders for materials. Here a separate prototype has been suggested for the creation of the purchase order, but the other two dialogues, dealing with different aspects of progressing the order, have been combined into one prototype, because, in the opinion of the designer, the two dialogues are relatively simple, and both are activities to be carried out by the same user.

When a large system is being developed, and a number of different prototypes are being constructed, it is necessary to provide some form of identity for each separate prototype. This is left to the discretion of the developer, but you will notice that in the examples provided here, a very simple naming convention is used.

2.2 Implementing Initial Prototypes

The purpose of this stage of the development process is to put forward and obtain agreement on designs for the format of the user/computer interface. This interface consists of

> the input forms and documents
> the output reports
> the dialogue between the user and the computer.

The process of deciding and agreeing these can bring to light new aspects of the system requirements. For instance, the recognition of the significance of some small piece of extra information to be included on an input document may force a reconsideration of part of the proposed system design, and even perhaps a re-assessment of some of the logical requirements. Should this happen, discussions on revision of the design must take place, and amendments must be made to the various models involved, as well as to any related supporting documentation. So this step in the design stage of the development is also an important part of the investigation and analysis of the system.

There are well established techniques for the design of input documents for batch-oriented computer systems, and the basic approach to the design of reports and other printouts from batch programs has been standard for many years. There will be some brief discussion of these later in this chapter. However, the main emphasis of the chapter is on the development of on-line dialogue, and the design of screen-based reports for on-line systems. It is suggested that this should usually be done by means of prototypes.

There are four basic activities involved in the process of implementing a prototype, and each of which will be examined in turn:

1. Discussing with the user the basic form of the dialogue

2. Building the prototype

3. Exercising the prototype

4. Obtaining agreement on its acceptance

2.2.1 Discussions with User

Before beginning to build a particular prototype, the analyst concerned will need to discuss with the proposed user the likely format of the dialogue. This discussion may be in the form of a walkthrough of the relevant part of the Systems DFD, followed by an exchange of views on format standards and preferences.

The relationship between the user and the prototype builder is very important, the success of the approach can depend on it. Milton Jenkins, a leading authority on prototyping, emphasises that ideally there should be a one-to-one relationship between user and builder, and that the user involved should be an intelligent, informed and responsible member of staff with interests beyond the limited area covered by the prototype. Often the most appropriate user is the supervisor of the person who will carry out the task when the system is complete.

2.2.2 The Building Process

The building of the prototype will involve the use of a modern fourth generation environment, which will include screen formatters or 'painters' to help design the input, and report generators to provide code for the main outputs. Often the prototyper is able to make use of pieces of other systems which carry out a similar task to the one required in the prototype, and tailor them quickly to fit the new requirements. Sometimes a 'template' system can be used as a starting point, thereby permitting a very fast development of the prototype.

2.2.2.1 Rapid or Evolutionary

The methods to be used in its production will depend on whether the prototype is to be 'rapid' (ie. to be built as quickly as possible and eventually thrown away), or 'evolutionary' (ie. to be used as a major part of the finished version). A decision on which of these two approaches to use must be taken before work on the prototype starts.

Sometimes it is obvious that the final system must be developed using different hardware and/or software. That being the case, only the 'rapid' approach would make sense. Examples of where this might happen include the situation when the development is for a mainframe and the 4GE being used is only available on a microcomputer, or when the required 'response time' for an on-line system is so critical that the 4GE cannot provide efficient enough code.

However, more and more often, the 'evolutionary' approach is being taken, and when this happens the structure of the prototype has to be much more carefully considered. Having said that, it must be pointed out that there are no well-recognised structured programming techniques for use with fourth generation environments. Structured program design is a method of identifying the best sequence for instructions to be obeyed within a program, and this implies that the programming language to be used is procedural (ie. the instructions indicate how actions are to be carried out and in what order). Fourth generation languages vary considerably in their formats, but almost all of them make use of non-procedural statements (ie. instructions saying what must be done, not how). Such statements are not necessarily obeyed in the same order in which they are written. As a result, traditional methods like flowcharting or JSP techniques are of little benefit.

One technique which has been found to be useful as a way of ascertaining the best structure for a 4GL routine is that of Logical Path Analysis. This is discussed in detail in chapter 12 as part of the analysis of data usage within the proposed system, where its contribution to the design of the complete database for the system is stressed. Basically, for each process, logical path analysis traces the sequence of accesses to the entity types containing the information needed to calculate and construct the process outputs. The diagrams used are of sufficiently high a level to relate to non-procedural

189

rather than procedural languages, though it must be admitted that the technique suits some languages better than others.

The suppliers of some 4GEs do recommend certain programming techniques which can be used to advantage with their particular product.

2.2.2.2 The Data Aspects

The prototyper should make use of the logical entity types identified during the business analysis, and should treat them as physical file-equivalents; (ie. each table becomes a separate file, with each entry being treated as a record). This is important, because the job of optimising and 'physicalising' the final database will normally be the responsibility of specialist staff. At this early prototyping stage those decisions can not normally be anticipated, so it is best to stay as closely as possible to the logical model. This is easier with some 4GEs than others, and the prototyper must obviously use some judgement.

A number of organisations who pursue a strongly data-oriented approach to systems development insist that the full logical data model must be completed before prototyping commences. There may often be good reasons for this, in that for the particular 4GE being used, once a physical database has been set up its structure is very difficult to change. However, in normal circumstances each individual prototype will deal with a small part of the data model, and this part can be readily identified (and even built) from the limited information available on the DFD and its supporting documentation. Nevertheless, the part of the data model to be used in the prototype must be agreed and confirmed with those responsible for the overall data modelling task.

2.2.2.3 Standards within the Organisation

In any well organised Systems Development organisation there should be a set of standards for dialogue design to which all analysts, including prototypers, must adhere. These standards will refer to two basic concepts, the dialogue style (ie. 'look and feel') and the use of common routines.

1. The 'dialogue style' for an organisation is expressed in the standards used for designing menus, commands and question & answer dialogues. For example, there should be a common method of menu selection (using either numbers or letters to select options). Similarly there should be common standards for 'help' and 'quit' commands. Menus themselves should be structured in an hierarchical pattern,

with a standard method of returning to the level above. The function keys should also have standard significance throughout the organisation.

2. There should be a library of common routines available to all prototypers. These routines will carry out tasks which might be needed in any number of prototypes, and they will normally have been specially written in as efficient code as possible; (some of them may actually have been written in 3GL). For example, there may be several used for the validation checking of input fields (eg. a check digit routine, or a date subtraction routine).

These standards, particularly those constituting the organisation style, actually impose limitations on the choices available to the users. If the standards are good, this should pose no hardship, and in fact should simplify and speed up the whole prototyping process.

2.2.2.4 Designing for Levels of Experience

One important problem faced by any dialogue designer is how to allow for different levels of knowledge and experience among users. A user beginning to make use of the proposed system may need to be given long clear messages for guidance and reassurance. However, when a level of experience with the software has been built up, those same helpful messages may no longer be necessary, and are likely to cause annoyance by slowing the user down.

Often, not only must two separate dialogue approaches be provided, one for the beginner and one for the experienced user, but there must also be a facility for graceful movement between the two. This is necessary because the beginner will gradually build up experience, and the experienced person will occasionally forget something basic. This means that the two dialogue approaches must be designed to interact.

2.2.3 Exercising the Prototype

Once the prototype has been built, the user must be shown how to use it: (some minor dialogue amendments may well come to light during training process). The prototype should then be left with the user for an agreed period of time, the length of this period being as little as a couple of hours, or as long as a week, depending on

the complexity of the prototype

the agreed deadline for the project
the user's workload on tasks other than the prototype.

Before the issue of the prototype to the user, agreement should be reached on the nature and scope of the checking activities to be carried out. The user should be encouraged to build up a series of computer records on the system (preferably taken from real-life cases), and work with them as though they were part of a live system. Extreme values (maximum and minimum) as well as error values for fields should be tested against the prototype, as should exception situations taken from the user's experience. All tests and conclusions should be formally noted down by the user.

At the end of the agreed period all the ideas put forward by the user for amendment and improvement should be given careful consideration. A revised version of the prototype can then be quickly developed, and returned to the user for further comment. This iterative process of adjusting the prototype to include user's suggestions can not however go on indefinitely! If this was allowed, there could be the danger of a pernickety user vaccilating between two shades of background colour of screen! Normally a limit of three iterations is set for any particular prototype or prototype stage, and a time period is agreed for each iteration.

2.2.4 Formal Prototype Acceptance

Once the prototype has been iterated and improved to the satisfaction of the analyst/prototyper and the user(s) involved in its exercise, it can then be put forward for formal acceptance as the completed product of a specific project task. This procedure is part of the project control of the development, and it allows the project manager to 'sign off' as complete, one of the items on the schedule. This does not mean however that the prototype is in its final form! It is quite likely that further versions of that same prototype will be required, and a greater level of functionality incorporated.

2.3 The Evolution of the Prototype

The kind of prototype we have been discussing here, and its two or three **iterations,** normally serve the purpose of settling the main dialogue and confirming the basic correctness of the approach. Such a prototype is mainly based on the logical business requirements, as identified in the analysis stage models, and it is usually referred to as the INITIAL prototype.

However, as the design process continues, more and more design detail becomes available for incorporation into the prototypes. This information comes not only from the user, who is examining the earlier prototype iterations, but also from the analyst/designer, who is probing for the more obscure requirements, and providing amended solutions.

The Systemscraft methodology attempts to elicit these extra design requirements in a structured way, using various modelling techniques to assist in the probing process, and decomposing the gradual development of the system into a series of clearly identifiable steps. This is dealt with in some detail in the next chapter, but the important point here is that additional analysis material from each of the steps can be used to construct new **versions** of the earlier prototypes, versions which contain more functionality and complexity, and are therefore much nearer to the complete form of the system which will eventually be delivered.

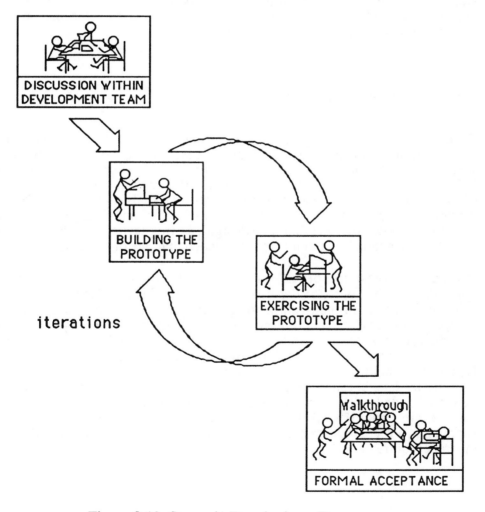

Figure 9.12 Stages in Developing a Prototype

So the Evolutionary approach involves the creation and issue of two different levels of prototype; it consists of a series of VERSIONS, and a number of ITERATIONS within each version. The release of the versions may coincide with the main steps in the structured design of the computer system, and this helps to enable the project manager to monitor the overall progress of the design and implementation of the system.

It should be stressed that not all prototypes will need a separate issue for each of the possible versions. One can progress through the development steps in a similar way to that used by the data analyst to ascend through the three 'normal forms'; ie. one may find that having completed a prototype version for one step, all aspects of the next step have already been satisfied, and the accepted prototype can be also be considered to represent that later step. Certainly, in the development of small systems, this versioning approach should not be allowed to cause unnecessary overhead, or to interfere with the creative flexibility of the designer.

2.4   Alternative Approaches

When a non-prototyping approach (sometimes referred to as 'pre-specification') is taken to the development of a system, the input forms and screens, and the output reports, will normally be designed using paper-based techniques. There is no shortage of material describing these activities; most of the standard texts on systems analysis provide detailed coverage of the subject. In the particular case of dialogue design, a number of methodologies provide dialogue design charting techniques (SSADM is an important example), and it is recommended that one of these be used.

Perhaps the most important difference between the non-prototyping approach and the approach described throughout this chapter relates to the level of formality in the documenting of the Process Descriptions for the computer processes. When no prototype is being built, the only record of the design has to be held in the models, and in the supporting documentation for those models. In particular, the text-based Process Description, held in some form of Structured English, must be rigorously kept up-to-date, because it represents the full specification of the process to be later handed to the programmer. By contrast, **any prototype is itself the basic specification of the process concerned**, and the supporting Process Description is needed only to simplify and clarify the inherent requirements.

10 DEVELOPING THE COMPUTER PROCESS DESIGN

The early steps in the Systems Design stage, as discussed in the previous chapter, present the user with a series of basic working models of parts of the proposed system. These models (prototypes) illustrate the dialogue which will take place between the user and the computer, and they provide, as a starting point, the simple functionality that the processes require. They also provide a mechanism (ie. the iteration) whereby the user can make adjustments to the dialogue and functionality, moving them gradually nearer and nearer to the full system requirement. Bearing in mind that these early prototypes are based on the product of a full Business Analysis stage, is this not itself a rigorous enough procedure to complete the process design? In many cases where the system being developed is simple and stand-alone, there may be little need for further design effort.

However, in the case of a medium or large system, or where the processes to be computerised are not obviously straightforward, then a more detailed and rigorous approach must be used, in order to guarantee that the full complexity of the system has been recognised and incorporated in the design.

In this chapter, three aspects of this extra complexity are examined, and modelling techniques to support them are illustrated. These aspects are:

1. Confirming the Process Detail; identifying and adjusting for the errors, the exceptions, the rare permutations of circumstance

2. Applying Necessary Controls to the system, to guarantee accuracy, security and privacy

3. Assembling the Computer System, combining the individual modules and prototypes into programs, suites, sub-systems, etc.

It is also proposed that the tackling of each of these three aspects be treated as a separate 'sub-step', a part of the larger design step of 'Developing the Computer System Design'. This breakdown should help the project manager of a large or medium-sized system to estimate and monitor progress more effectively. Figure 10.1 (which is an excerpt from Figure 3.3) illustrates these sub-processes and the main modelling techniques recommended for carrying them out.

The sequence in which these sub-steps are to be performed may depend on the overall development approach being taken. If for example a traditional batch-oriented 'pre-specification' design method is being employed, then there are good project management reasons for constructing the programs, suites and sub-systems as early as possible; these components are used as basic units of work, and the progress of the project is measured in terms of their completion.

On the other hand, if evolutionary prototyping is being used, the process of combining the modules can sensibly be left until all other complexities of the computer processes have been sorted out. There is also some justification for leaving the detailed examination of controls until the completeness of the processes has been examined; the completeness check itself often brings to the surface situations where extra controls are required.

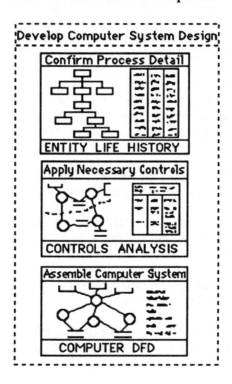

Figure 10.1 Components of the 'Develop Computer System Design'

Again, in a medium-to-large system, these sub-steps can be used to separate out different 'versions' of the evolutionary prototypes (as suggested in the previous chapter). However, in small developments there may be little advantage in breaking the higher-level step down, and all three sub-steps may be undertaken as one, with the recommended techniques cut back to a minimum.

There is one important difference between 'Develop the Computer System Design' and the earlier analysis and design steps already discussed. Whereas until now we have been developing individual parts of the system, as integral units (eg. sub-functions and prototypes), the techniques used from now on examine the system 'across its breadth'. This means that the full Business Requirements Specification (especially the entity and relational model) and the Systems DFD for the complete system must be available as a pre-requisite for this step. Again, this has critical project management implications.

1   CONFIRMING THE COMPUTER PROCESS DETAIL

It is generally accepted that the most important technique used for the identification and development of processes (whether business or computer) is the Data Flow Diagram. The DFD has notable strengths and advantages, many of which have been demonstrated in earlier chapters, but there are a number of aspects relating to the analysis and design of processes for which this modelling technique is inadequate. To describe these as 'weaknesses' may be a little unfair; the technique is not claimed as a complete stand-alone method of analysis and design.

It has long been recognised that there are three major areas of process design for which the DFD does not provide full support. These are:

1. The recording of volumes, frequencies, and trends; on a DFD there is no way to show how many occurrences of the activities or data the system is required to handle.

2. The sequence of events and activities; although the DFD processes sometimes have an implicit order, in that the output from one process may be the input to another, there is no formal way to show the time element involved between processes.

3. A completeness-checking facility; because the DFD technique provides only a relatively 'fuzzy' method of process definition (in one sense one of its main strengths!), it is not possible to use the technique in a formal way to 'prove' its own completeness. It is not unusual for a junior analyst to overlook an important functional requirement while building a DFD, and only through a comparison with views provided by other forms of model does the omission come to light.

The first of these missing views of the system, that relating to the numerical aspect of processes, is covered by techniques discussed in chapter 11, 'Analysing Data Usage'. It is suggested here that the other two areas not supported by the DFD can best be covered by using a technique which is common to many of the best known and most effective methodologies, a technique known as Entity Life History Modelling.

1.1 Entity Life History Modelling

Entity Life History modelling involves treating the proposed computer system as being made up of more than just functions and data. If one were to attempt to define an information process in simple algebraic terms, the definition might look like this:

'When A occurs, do B to C'.

In such a definition, the 'B' variable is clearly an activity or FUNCTION, and the 'C' variable is equally obviously DATA, but what does the third variable 'A' stand for?

The answer is an EVENT. An Entity Life History model is produced for each entity type in the system, and it identifies all the EVENTS which will have an effect on an entity during its life in the system. The events identified are then checked against the DFD processes to make sure that the full life of the entity, from its creation to its deletion, are covered somewhere in the proposed computer system.

So an entity life history is a model of the BEHAVIOUR of an entity from the time at which it becomes of interest to the system to the time when it ceases to be of interest. The model plots the series of events that can happen to the entity, showing their order and dependency. It highlights situations where missing events may cause an error, and where other events occurring out of sequence may require complex processing for recovery.

There are in fact several different forms which an ELH model can take, depending on the level of complexity of the activity affecting the entity. The simplest form involves listing the events and the DFD computer processes in which they occur (eg. Figure 10.2), whereas the more complex forms comprise a hierarchical decomposition model showing the different possible sequences of events (eg. Figures 10.3 and 10.4). These different forms and the symbols used are explained as the chapter unfolds.

The description of the ELH technique and how to use it is covered under the following headings:

1. Events and their Effects

2. Levels of Complexity of ELH Models

3. The ELH Modelling Symbols

4. Building an ELH Model

5. Applying State Indicators

6. Incorporating ELH Results in the Design.

ENTITY TYPE	EVENT /EFFECT	SYSTEMS DFD PROCESS	REF .
	Customer Created	Validate Customer	2.1.1
	Name & Address Changed	Handle Enquiries	2.2.1
CUSTOMER	Credit Status Changed	ACCOUNTS SYSTEM	—
	Customer Deleted	Delete Obsolete Customer	5.3.1

Figure 10.2 Example of the Simple Form of Entity Life History

1.2   Events and their Effects

As already stated, the building of an ELH involves examining each entity type in the system, and identifying for each in turn all the events which affect it. This relationship between events and their effects is important.

An event can be one of three different types:

1.   An EXTERNAL event, caused by something happening outside the system: (eg. a customer placing an order).

2.   An INTERNAL event, caused by some other process within the system: (eg. an item of stock going below its re-order level, creating the need for a purchase order).

3.   A TIME-BASED event, one which must occur either at a fixed time (eg. end-of-the-week payroll), or after a period of time has passed (eg. a reminder sent two weeks after the invoice).

Similarly, the effect of an event on an entity table can be one of only three types:

1. It can CREATE an entity

2. It can DELETE an entity

3. It can CHANGE the value of one or more of the entity's non-key attributes.

So, only those events that cause the contents of the entity table to be altered are shown in an entity life history model.

The fact that we are interested in both the event and the effect of the event on the entity is reflected in the naming conventions used. As far as the ELH builder is concerned, the event and its effect can be thought of as happening simultaneously, so the name given to each event/effect on the model should as far as possible reflect both.

For example, in the case of the ELH model for the Customer (Figure 10.2), the event which causes the creation of a new customer entity is the arrival of the first order for that customer. The name chosen here, 'Customer Created', emphasises the effect rather than the event.

On the other hand, in the case of the 'Materials' ELH (Figure 10.3), a higher-level event name emphasises the effect (eg. 'Stock Balance Incremented'), and the lower-level names identify the different events which have that same effect (ie. 'Materials Delivered' and 'Materials Returned by Production').

It should be noted that an event/effect shown on an ELH diagram is NOT the same as a DFD process: whereas the process describes what must be done, the event/effect simply indicates what happens. This difference is reflected in the naming conventions used for each:

A DFD process name is in verb/object form, and in imperative mode (eg. 'Take Order', 'Inspect Materials').

An ELH event/effect is often in object/verb form, is passive, and simply indicates what occurs (eg. 'Order Created', 'Materials Accepted', etc.)

It is essential that standards like these are used, so that a clear distinction between the components of the two models can be made.

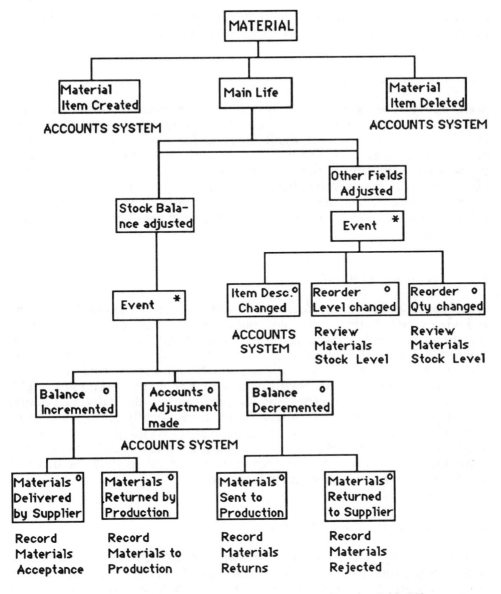

Figure 10.3 Example of the Standard Form of Entity Life History

1.3 Levels of Complexity of Entity Life History Models

The ELH modelling technique enables the designer to examine in great detail ALL possible events relating to ANY of the entities held within the entity table. Events can occur within a normal sequence, requiring standard procedures to be carried out, or they can occur in the wrong sequence, causing complex error procedures to be set in motion.

1.3.1 The Simple Form

Some entity types are effected by only a limited number of events which can occur in a very simple sequence. Often the fact that events may occur in different sequences is not significant and does not give rise to complex error procedure coding. Entity types like this (for example 'link' entity types) may only require the simplest form of ELH model, as shown in Figure 10.2. In this example, the 'Customer' entity is seen to be subjected to four possible events, the sequence of which is fairly straightforward. All that is necessary is to list the events, and ascertain to which DFD processes they relate.

1.3.2 The Standard Form

When the order of events effecting the entity type forms a complex pattern, and where the occurrence of events out of sequence can cause problems or signify error, then a more rigorous form of the ELH model must be employed. Figure 10.3 gives an example, and shows how the model for the 'Goods' or 'Stock' entity type might look. Notice how the life history is shown as a series of events occurring in a time sequence from left to right, starting with the birth and finishing with the death of the individual entity. The main body of the life is broken down hierarchically into groupings of event/effects, and then into events themselves. Everything that can happen to any of the Goods entities is shown in this diagram. Finally, under each individual event (shown as a bottom-level box on the model), the name of the DFD process in which it occurs is placed.

1.3.3 The 'Status Indicator' Form

Where the life cycle of the entity type is very complex, and there is a need to be able at any time to indicate the current 'status' of each entity, then a more advanced form of the model is used. It implies that an extra attribute is to be added to the entity type definition, an attribute known as a status indicator. The value in this attribute will be examined whenever an event occurs, and its purpose is to enable the system to identify whether the event is occurring at a valid time in the entity's life. For example in Figure 10.4, which shows the ELH model for the 'Order Line' entity type, the event/effect 'Line Cancelled' must not be allowed to happen if any of the items in the line have been delivered; (clearly this could destroy the audit trail!). In the model, a number is allocated for each type of event, and the status indicator field is updated with that value when the event/effect takes place. These numbers are shown under the appropriate boxes on the diagram. In Figure 10.4, a value of '4' is placed in the status indicator when any part of the order line is delivered. So, if when there is a request to cancel the order line the indicator already has a value of '4' or more, then the request is invalid, and the line must not be cancelled. (This is explained in much more detail later in the chapter.)

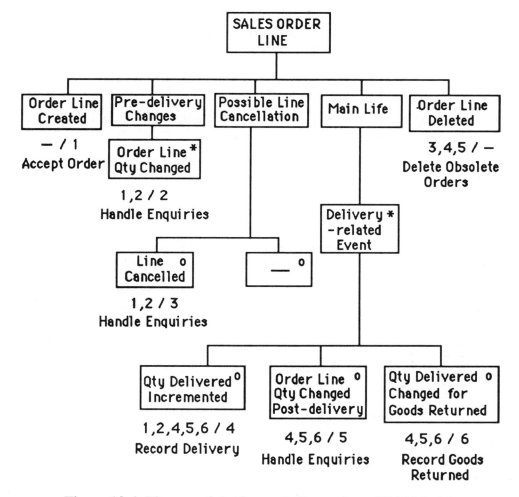

Figure 10.4 The use of the Status Indicator in an ELH Model

Although the designer may choose any one of these three forms of the model for entity types within the system, it is necessary also to provide an overall view of the system as shown by the entity life history model. This can be done by reducing all ELHs to the simple form, and recording them in one document, (see Figure 10.5).

ENTITY TYPE	EVENT / EFFECT	SYSTEMS DFD PROCESS	REF
	Material Item Created	ACCOUNTS SYSTEM.	—
	Materials Delivered by Supplier	Record Materials Acceptance	3.1.1
	Materials Returned by Production	Record Materials Returned	3.2.1
	Accounts Adjustment made	ACCOUNTS SYSTEM	—
MATERIALS	Materials sent to Production	Record Materials to Production	2.4.1
	Materials Returned to Supplier	Record Materials Rejected	2.4.2
	Item Description Changed	ACCOUNTS SYSTEM	—
	Re-order Level Changed	Review Stock Levels	2.3.3
	Re-order Quantity: Changed	Review Stock Levels	2.3.3
SALES ORDER LINE	Order Line Created	Accept Order	1.3.4
	Order Line Cancelled	Handle Enquiries	1.3.6

Figure 10.5 The Combined ELH Model

1.4 The ELH Modelling Symbols

The standard form of the entity life history model always begins with the name of the entity type in a box at the top of the diagram. Beneath this there are a number of boxes, each representing either an individual event/effect or a grouping of such events. They are arranged in a hierarchy, the levels of which are defined by the lines connecting the boxes, and there are a number of 'constructs' or conventions used to indicate the order in which event/effects can occur. These constructs enable the ELH builder to represent the following aspects of the entity life:

1. Sequence (where events must occur one after the other).

2. Selection (where several alternative events may occur).

3. Iteration (where the same event or group of events can occur a number of times).

Two further conventions are provided, enabling the modelling technique to handle all possible combinations of circumstance. These are:

4. Parallel Structures (where groups of events are unrelated in terms of order).

5. Quit and Resume (where the order of events can be altered by special circumstances).

Each of these constructs and conventions will be discussed in turn.

1.4.1 Sequence

The fact that a number of event/effects must occur in a fixed sequence is indicated on the ELH model by placing the event names in boxes at the same hierarchical level, and arranged in order of occurrence from left to right.

It should be noted that there are no symbols enclosed within the event boxes; only the chosen event name is present. Figure 10.6 gives an example of what is probably the most common use of the sequence construct, the first level of the hierarchy, showing the birth, main life, and deletion of an entity. Whereas the 'creation' and 'deletion' boxes are likely to represent individual events (and therefore can not be sub-divided), the 'main life' box represents a group of events, and will need to be expanded at a lower level.

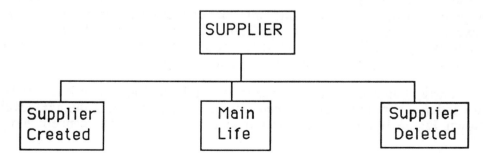

Figure 10.6 An Example of ELH Sequence

205

1.4.2 Selection

The situation where one of a number of alternative events might occur is indicated on the model by a box for each of the alternatives, and a special symbol 'O' placed in the top right-hand corner of all such boxes.

The example provided in Figure 10.7 shows the high-level sequence discussed earlier, and a selection of events at the next level. Note that although there are only four possible events in this entity life, they can not be shown at the same hierarchical level, because some occur in sequence and some are alternatives. ALL EVENTS AT THE SAME LEVEL OF A HIERARCHY LEG MUST BE OF THE SAME CONSTRUCT. In order to get round this problem, a 'group' box is placed in the sequence ('Main Life'), and that can then be split into a selection.

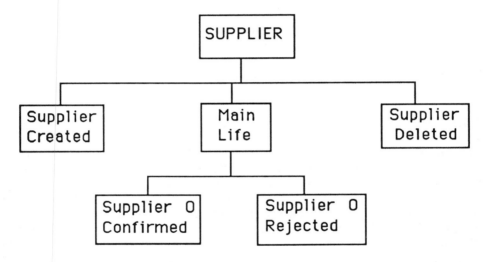

Figure 10.7 Example of an ELH Sequence and Selection

The term 'selection', as used in the ELH modelling process, has a specific meaning. Where a number of alternative events are shown on the model as a selection, it means that one and only one of them can occur (ie. they are 'exclusive' alternatives). Also, it indicates that one of them MUST occur. If there is a possibility that none of the events will occur, then a 'null' box (with a dash in it) must be provided as one of the alternatives.

Similarly, if an event in a sequence is seen to be optional, then it should be treated as a simple selection (occurring or not occurring), at the next level down. Figure 10.8 illustrates this with an example of an order cancellation.

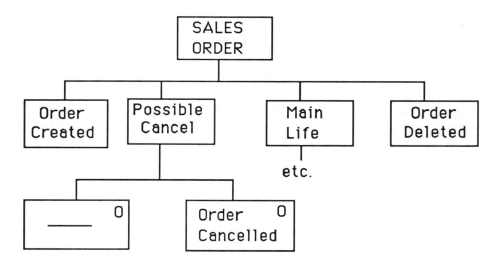

Figure 10.8 An Order May or May Not be Cancelled

1.4.3 Iteration

The Iteration construct allows the ELH builder to describe events or event-groups which are to be repeated a number of times within the life of the entity. For example, in the life of a 'Stock' entity, the Stock Balance value will be adjusted every time an ordering event occurs for that particular type of stock. Similarly, in the case of a 'Customer' entity, the Credit Status value may be changed many times by the same kind of event (late payment). An event which is repeated must be shown ON A LEVEL OF ITS OWN in the diagram, and an asterisk symbol is placed in the top right-hand corner. Figure 10.9 gives the example of 'Customer' entity type discussed above. The other example mentioned, that of the 'Stock' entity type, is shown in some detail in Figure 10.3.

Again, the term 'Iteration' has a specific meaning within the ELH modelling process. Although it can be said that an iteration is an event or event-group which is repeated many times within the life, the word 'many' is taken to mean

               zero,
               one, or
               more than one.

This means that an event shown as an iteration may not occur at all for some of the entities in the table!

For example, in Figure 10.9, some customers may never have any changes to their name and address, or any adjustments to their credit status.

207

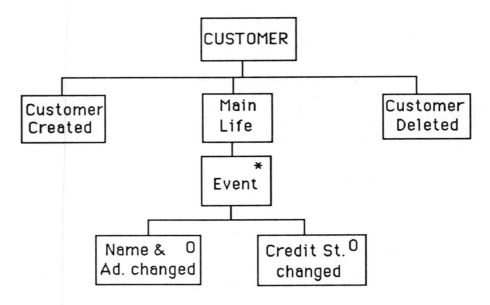

Figure 10.9 An ELH for the Customer Entity Type

A combination of the selection and iteration can be used effectively to simplify the order of events within a life. Figures 10.10 and 10.11 illustrate this.

Figure 10.10 shows the life for the 'Invoice' entity type (which is part of the Gentry Accounts system). As can be seen, this is a complex life, with a strict ordering of events. The invoice is created, then there is a period during which some payment(s) may be made. If the full amount has not been paid by the due date, a reminder is sent. This may trigger a series of payments for the outstanding amount. Eventually, if it becomes obvious that the customer does not intend to pay, then action is taken to record a bad debt. When, several years later, the invoice becomes so old as to be no longer legally required, it is deleted.

On the other hand, Figure 10.11 shows a much simpler version of the same thing. This model describes the Invoice entity life as a creation event and a deletion event, with the main body of the life being made up of a series of events, either payment, reminder or bad debt action.

Which of these is correct?

Clearly both of them are correct, but one contains more information than the other: the natural sequence of the main events has been ignored in the second model. This is not a problem in the example quoted; both models

208

can be checked against the Systems Data Flow Diagram, and will probably give the same result.

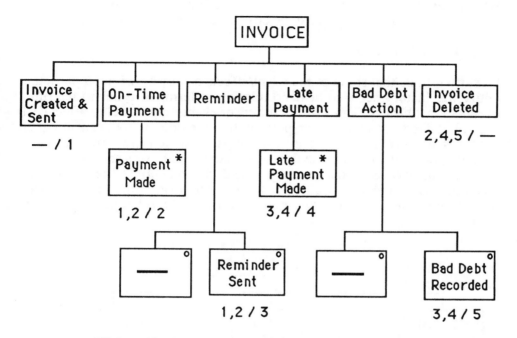

Figure 10.10 An ELH for the Invoice Entity Type

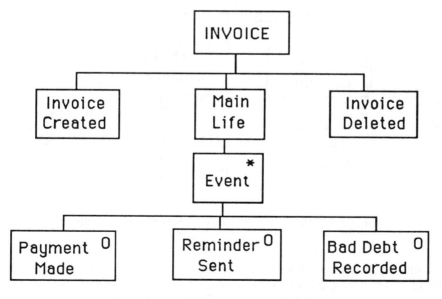

Figure 10.11 A More Basic ELH Model of the Invoice

It is however recommended that the ELH builder begins by trying to produce the more complex life, because very often it is the building process which identifies those events which seem at first to have little significance, but on closer inspection prove to be critical to the system's success.

1.4.4 Parallel Structures

One often finds that some of the event/effects occurring within a particular entity life history seem to bear no relation in terms of order to the other events in that history. These different sets of events appear not to affect each other, but can overlap and intersperse within the time period, causing great numbers of permutations of events, most of which are irrelevant. For example, the updating of static fields like names, addresses and descriptions can often be allowed to occur at any time during an entity's life. This could mean that such an event would have to be shown many times on the model, placed between all of the other events showing the main progress of the entity.

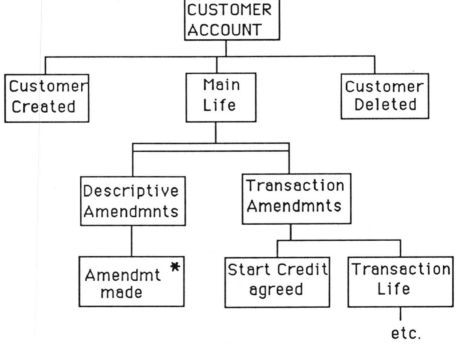

etc.

Figure 10.12 A Parallel Structure Example

To avoid this, a convention called a 'Parallel Structure' is employed. This is in the form of a double line, and it separates off different event/effect groups which can occur simultaneously. In effect, it provides the opportunity to model more than one 'life' for the same entity in the same diagram.

Figure 10.12 gives an example relating to the 'Customer Account' entity type (another entity type from the Gentry Accounts system). The parallel structure separates off the events which update the basic descriptive fields of the entity and those which carry out the account transactions. A further example of the parallel structure occurs in the earlier Figure 10.3, where the events which update the stock balance are separate from those which amend the other attributes.

1.4.5 Quit and Resume

Some of the events which can occur during an entity life are rare and exceptional. They may even cause a major change to the normal order in which other events are likely to take place. For example, if a Sales Order is cancelled shortly after it has been set up, then the order entity will be marked as cancelled, and all the remaining events in its life (with the exception of the deletion event) will no longer be carried out.

The 'Quit and Resume' convention allows such a change of event sequence to be modelled. At the point where the event causing the change occurs, the letter 'Q' is placed: (this represents the quitting point). At the point where the interrupted life is to resume, the letter 'R' is placed.

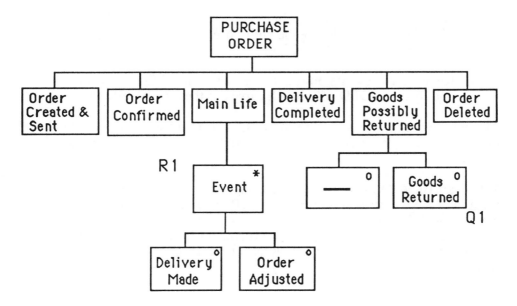

Figure 10.13 The Purchase Order ELH Model

The example shown in Figure 10.13 illustrates the opposite kind of situation, where the natural life of the entity is almost complete. The model for the 'Purchase Order' entity type shows a situation where the complete

211

order has been delivered, but after inspection, some materials are found to be unsatisfactory and are returned to the supplier. At this point, the life cycle of the entity must resume at a point before the completion of the delivery.

On rare occasions it is necessary to allow for more than one 'Quit and Resume' case in the same entity model. When this happens, the 'Q' and 'R' are given subscript to relate them: (ie. the first case is marked 'Q1' and 'R1', the second is marked 'Q2' and 'R2', etc.).

1.5 Building an ELH Model

There are four basic stages in the process of building an ELH model. They are:

1. Examining the ATTRIBUTES of the entity type, and identifying the potential value range of each.

2. Identifying and listing all EVENTS which can cause a change to occur to one or more of those attribute values.

3. Assembling the events in the appropriate ORDER OF OCCURRENCE during the entity life, and recording them in diagramatic form (the model).

4. Relating the events on the model with the PROCESSES on the Systems DFD which will carry out the events' effects.

The whole activity is a 'design' process, and requires both analytical and creative thought; new events, attributes, and processes may come to light as a result.

When an ELH has been completed, the findings must be disseminated. Details of the events and their required effects must be passed to those who are responsible for the detailed design of the computer system processes. These details can then be used to expand the process specification, and perhaps to provide a NEW VERSION for the PROTOTYPE of which that process is a part.

This stage in the development of the ELH model is particularly important. It involves cross-checking the design requirements found using one major modelling technique with those found using another, and can illustrate the need to revise aspects of either or both models.

There are three types of situation which, when identified here, may highlight the need for further action:

1. Missing Processes. Sometimes no DFD process can be found for a particular event on the model; (this is quite common for example in the case of 'deletion' events). When it happens, the completeness of the proposed computer system is called into question, and the Systems DFD is re-examined and adjusted.

2. External Events. Some of the events recorded on the ELH model may be dealt with outside the system under consideration. They are performed as part of some other computer system, either already in existence, or being developed separately. Such interfaces with other computer systems need to be checked very carefully with the analysts responsible.

3. Duplicate Events. It may be that a particular event on the ELH is found to occur in more than one DFD process. Whilst this is not necessarily an error, it does suggest the possibility of duplication within the systems design, and this aspect should be considered.

1.6   Applying Status Indicators

Earlier in the chapter, it was suggested that where the entity life cycle was complex, and where events might occur in the wrong order, thereby enabling errors to occur, a more advanced form of the ELH model should be used.

This approach involves the addition of an extra attribute known as the 'STATUS INDICATOR' to the entity type, and the setting of a value for this status when each event occurs. In computer processing terms, this means that when an event effecting an entity takes place, the existing status of that entity can be examined, and checks can be made as to whether the new event can be accepted as valid at that time.

For example, when a delivery event is notified, the status of the appropriate 'sales order line' entity will be examined.

> If that status shows that the order line has already been delivered, then some form of error (perhaps a duplicate delivery) has taken place.

Similarly, if the status shows that the order line has been cancelled, a different kind of error procedure is necessary.

1.6.1 The Symbols Used

As well as the extra attribute, one further modelling convention is made use of in the advanced form of the ELH model. This consists of a series of numbers placed under each event box, and it takes the form:

$$x / y$$

where   x   is a list of status values which are valid when the event occurs

            y   is the value to which the status indicator is to be changed.

A simple example is given in Figure 10.14. Here the event in the box will only be valid if the status indicator is already set to values 1 or 2, and the event itself should cause the status indicator to be set to a value of 3.

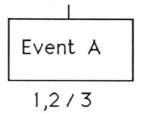

1,2 / 3

Figure 10.14 Status Indicator Value Symbols

1.6.2 Building the Advanced Model

Status indicator values are added to the ELH model after the normal building process has been completed;

1. The first event on the model will be the creation of the entity. Obviously there is no prior value to the status indicator, so 'x' is shown as a dash.

2. The values allocated to the event boxes are simple integers. These numbers have no significance other than to provide a unique identifier to the status-causing event (ie. they do not need to be in ascending order through the life).

3. The final event causes the deletion of the entity, so there can be no new value allocated to the status indicator. Again, a dash is used.

4. Not all events need to be allocated a status value; only where the occurrence of the event affects the validity or otherwise of other events in the cycle should a status number be given. For example, a change to a descriptive field like a supplier's address is unlikely to affect the entity's status, so can be left un-numbered.

The 'Sales Order' example is an ideal candidate for the use of a status indicator. Several of the important events in the life of an order (like 'cancellation' and 'completion') do not cause any change to the existing attributes, so there is clearly a case for some extra field to indicate the status.

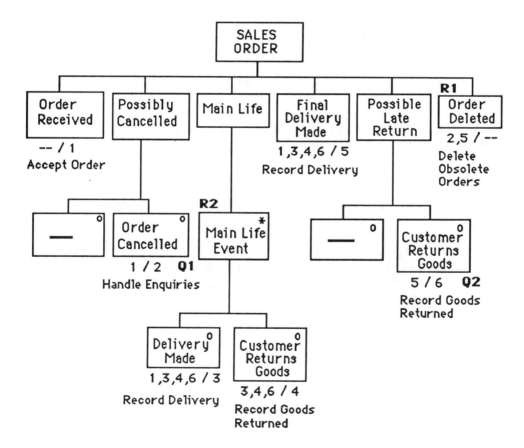

Figure 10.15 Sales Order ELH with Status Values

215

Figure 10.15 illustrates how status values can be allocated to the events in the 'Sales Order' ELH. The following points are worth mentioning:

1. When the order is cancelled, a status value of 2 is set. This means that no further activity can take place on this entity other than its deletion. The deletion event itself can only take place if an order has been either completed (4), or cancelled (2).

2. The final delivery event may be either the one-and-only delivery (previous status 1), or the last in a series of deliveries (previous status 3). Very occasionally this delivery may follow the late return of goods by a customer (status 5).

One final point about the building of ELH models should be noted. Although individual entity types each have a separate model, there are close relationships between some entity types, and design decisions affecting one may equally affect another.

For example, events in the 'Sales Order Line' ELH (Figure 10.4) are clearly related to the events in the 'Sales Order' ELH that we have just built (Figure 10.15). The same is equally true of the 'Delivery' and 'Delivery Line' ELHs.

Because of this, the design of an ELH for a particular entity type must be cross-checked and co-ordinated with designs for related entity types.

1.7 Incorporating the Results in the Process Design

The whole purpose of building a set of ELH models is to ensure the completeness of the design for the computer processes in the new system. So, the findings and decisions resulting from the model-building process must somehow be incorporated in the individual computer process design specifications. This can involve a number of activities.

1.7.1 Dealing with Un-allocated Events

During the construction of the ELH models, some events may be identified which are not part of a proposed computer process. At the time, the event box concerned may simply be marked with a question mark instead of the appropriate process name. Later however, the whole System DFD model needs to be re-considered, and a place found for the missing event. This may involve the creation of a new computer process, or the revised definition of one already in existence.

1.7.2 Checking for Un-modelled Processes

When the complete set of ELH models has been built, it is necessary to check that all the processes on the Systems DFD which affect any of the entity types have been taken into account. If a process has been overlooked, it may mean that there is some duplication in the DFD, or that certain events are missing from the ELH. In either case, revisions must be made to the appropriate model.

1.7.3. Disseminating ELH Findings to the Process Designers

The products of the ELH modelling activity must be made available to the analysts responsible for the development of the individual systems DFD processes involved. They will need this information in order to be able to expand the process specification (see Figure 10.16), and/or to provide a more detailed version of the prototype for user examination.

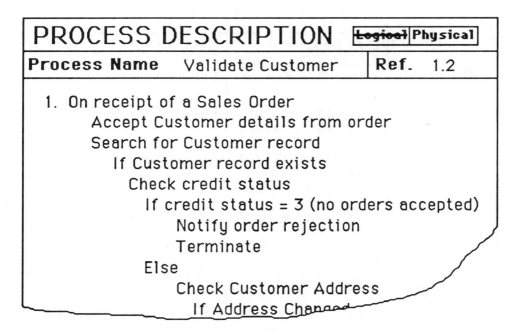

Figure 10.16 An Expanded Process Description (Post-ELH)

It is suggested that the 'walkthrough' is probably the most useful device for disseminating the detailed information; the process designer may have to make judgements on which error situations to allow for and report on, and to do this it is necessary to understand some of the thinking behind the model.

ENTITY TYPE	EVENT / EFFECT	SYSTEMS DFD PROCESS	REF
SALES ORDER.	Order Received	Accept Order	2.1.1
	Order Cancelled	Handle Enquiries	2.2.1
	Delivery Made	Record Delivery	4.1.2
	Final Delivery	Record Delivery	4.1.2
	Late Return Made	Record Returns	4.4.1
	Order Deleted	Delete Old Orders	4.3.2
SALES ORDER LINE.	Order Line Created	Accept Order	2.1.1
	Line Cancelled	Handle Enquiries	2.2.1
	Order Line Changed	Handle Enquiries	2.2.1
	Quantity Delivered Incremented	Record Delivery	4.1.2
	Quantity Delivered changed as Goods Returned	Record Returns	4.4.1
	Order Line Deleted	Delete Old Orders	4.3.2
CUSTOMER	Customer Created	Validate Customer	2.1.3
	N & A Changed	Handle Enquiries	2.2.1
	Credit Status Change	ACCOUNTS SYSTEM	—

Figure 10.17 A Combined ELH Highlighting a Process

Not only should full copies of the relevant ELH models be provided to the process designers, but there should also be some overview of which events effect the same processes. One of the best ways of showing this is by using the type of model referred to earlier as the 'Combined ELH', and illustrated in Figure 10.17.

In projects where an evolutionary development approach is being taken, most of the computer processes now being revised and expanded will have already been built as initial prototypes, and some level of agreement may have been reached with the user.

It is recommended that all the amendments from different ELHs which relate to the same prototype should be grouped together and issued as a new version of that prototype. Such a version is known as a COMPLETENESS prototype, and the use of this tactic can minimise the problems of project control inherent in an iterative approach.

1.7.4 Notifying the Database Designers of Data Changes

Obviously the analysts responsible for the design of the database for the new system must be informed of the extra 'status' attributes, and any other changes to the data model resulting from the ELH modelling process. It should be noted that the introduction of status indicators and other devices to simplify the processing moves the data model away from the purely 'logical' concept emphasised during the Business Analysis stage, and towards the physical implementation of the model as a database.

1.8   Summary

So, it can be seen that the Entity Life History modelling activity results in the revision and expansion of both the computer process definitions and the database specification.

This has been a relatively brief description of what can be a complex and rigorous modelling technique. Readers who are new to the technique and wish for further information about it should consult one of the many books which deal with either the SSADM or JSD methodologies (though it should be stressed that the approach they take to ELHs is slightly different to that taken here).

2 APPLYING NECESSARY CONTROLS TO THE SYSTEM

At some stage in the development of any project, a thorough, formal examination must be conducted into the controls necessary for the correct performance of the proposed system. This of course applies to the whole business system, not just the computerised information system, and the controls analysis study should involve not only the systems analysts but the user managers, the internal auditors and the person responsible for project management. The subject of controls is often thought of as very dull, involving the rigorous adherence to a series of checklists while plodding laboriously through the system detail conscientiously plugging the security gaps. On the contrary, if the correct approach and attitude are taken, it can prove to be one of the most interesting stages in the project, where the analyst has to put himself in the mind of a possible master criminal, and attempt to anticipate and pre-empt his every move!

But that is only part of the picture. There are three main aspects of the system which need to be protected by controls; they are:

> **ACCURACY.** Checks must be made that the transactions being carried out are performed accurately, and the information being held within the company's database is correct.

> **SECURITY.** There is an over-riding requirement to safeguard the assets of the company, making sure that no losses occur, either through omission or commission, whether deliberate or accidental.

> **PRIVACY.** There is also a need to check that the rights of individuals and other companies are protected. Perhaps the most important aspect of this is to make sure that the proposed system adheres to the restrictions laid down by the Data Protection Act.

The investigation and analysis of controls requirements are carried out by following a CONTROLS ANALYSIS TECHNIQUE, which is explained in detail in the rest of this chapter. The technique also suggests how the design of the controls should be carried out. The incorporation of these new controls into the already prototyped system should be done by issuing a new version of each prototype, the CONTROLS PROTOTYPE. This should be pushed through the same procedure as the earlier versions, checking the soundness and user-acceptability of the new components.

Experienced systems analysts will have at their disposal a massive range of potential solutions to all kinds of controls problems. For example, in the area of business controls, the analyst may wish to suggest the 'division of labour' as a safeguard in an accounts area. Similarly, a particular 'span of control' may be the optimum for the supervision of staff carrying out certain tasks.

Within the area of systems controls, the analyst may suggest the use of passwords to prevent unauthorised access to the system. In a batch system it may be necessary to incorporate a number of batch control devices such as headers and trailers, whereas for an on-line system, special file protection techniques such as 'before and after imaging' may need to be considered. Critical input data such as Account numbers may be considered to need the extra security of a check-digit , while some types of output may require pre-numbered pre-printed stationery.

The important point to make here is that this book does NOT attempt to cover the different forms of control option available. This is handled in admirable detail elsewhere in existing systems analysis literature. An assumption is made that the analyst is already aware of these options, and the purpose of the chapter is to provide a technique whereby the most appropriate controls can be applied where they are needed.

2.1 The Controls Analysis Technique

The controls analysis technique suggested for this methodology is a simplified version of an approach used by a number of the major multi-national oil companies. It is based on the use of the Data Flow Diagram, and it involves navigating round the various DFD models, identifying the origins of any controls weaknesses. In this chapter we will concentrate on the controls related to the computer system, so the model to be used will be the Systems DFD.

The C.A.T. approach is made up of a number of stages. They are:

1. **Identifying Exposure Points within the System.**
   Exposure Points are points at which information belonging to the company is potentially accessible by people inside or outside the organisation. This refers not only to the obvious forms of output, like purchase orders and invoices, but to any information within the company, which, if mis-used, may put assets at some risk.

For each individual exposure point identified, the following three activities are carried out:

2. **Identifying the Types of Threat from Exposure.** These types of threat include deliberate actions such as theft or vandalism, but also comprise risks to company assets and loss of business as a result of, for example, poor management decisions. The level of the threat in terms of the potential damage to the company is also considered and calculated.

3. **Identifying the Threat Situations.** Having looked at the possible threats, the team can then attempt to examine how those threats can arise. This involves using the DFD model, and tracing back from the exposure point, examining the circumstances represented by each process and the potential error from each flow. This stage of the controls analysis requires a great deal of imagination and creativity. One other aspect examined at this stage is the probability of the threat situation occurring. This information, together with the earlier details of the 'level of threat' enables the controls team to decide on the importance of the danger, and helps them decide on the extent of control to be exercised.

4. **Designing the Required Controls.** Having ascertained the extent of potential damage that can result from the exposure, the designer has to decide how to apply physical controls to prevent or minimise this damage.

CONTROLS ANALYSIS FORM

SYSTEM NAME
Materials Proc.

OUTPUT / FILE NAME	**Materials Inventory**				

PURPOSE
Maintains details of current material stock levels etc.

CONTENT/DESCRIPTION
Stock No., Stock Description, Quantity-in-stock, Re-order Level, Re-order Quantity

THREAT	DANGER H/M/L	SITUATION	DFD REF	PROB. H/M/L	CONTROLS

Figure 10.18 The Controls Analysis Form

The team of people conducting the Controls Analysis exercise will each have their own special areas of knowledge, and the quality of the exercise will depend on how well the individual contributions can be brought together. Perhaps the most effective tool to assist in this is the Controls Analysis Form (Figure 10.18), which acts as a kind of checklist, and is completed for each separate exposure point. It helps the team work their way through the exercise, stage by stage, and serves as the documentation of the results.

2.1.1 The Exposure Points

Identifying the exposure points is a relatively simple activity. Basically any data flow on the DFD which goes to an external agent represents an information output, and as such is an exposure point.

If the team is examining the narrower field of just the computerised information system, then any data flow passing from the computer to the user part of the system is deemed to be an exposure point. It should be noted that many of these may in fact be input-outputs, and may represent computer-user dialogues. This type of exposure point is of particular importance, as it is potentially one of the most dangerous.

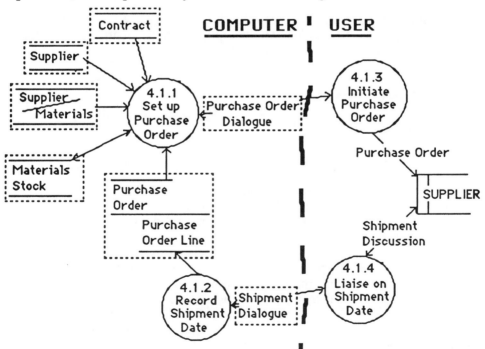

Figure 10.19   Exposure Points on a Systems Data Flow Diagram

One further type of exposure point other than a direct output can be identified from the Data Flow Diagram. This is the DATA STORE or file. Even though there may be no direct access to a particular file in the proposed system, it may possibly be subject to some form of un-authorised access, or may suffer accidental corruption. Clearly this could cause the company some damage, and therefore must be guarded against.

Figure 10.19 shows a Systems DFD, and the dotted rectangles encompass the exposure points identified on it. For each of these a Controls Analysis Form is set up, and the heading details completed. Figure 10.18 illustrates the heading details for one of these exposure points, the Materials Inventory file. (Obviously, although the same 'file' exposure points may occur on several different DFDs, it is only necessary to create one Controls Analysis form per file.)

Once all of the exposure points have been identified, and a form set up for each, then the next stage of the controls analysis exercise can be conducted.

2.1.2 Threats from Exposure

Each of the exposure points is examined carefully, based on the content of the information that it provides. An attempt is made to identify each potential THREAT that can result from this exposure.

2.1.2.1 Types of Threat

There are five basic types of threat:

>   Theft
>   Loss of assets
>   Mis-informed decision
>   Excessive cost incurrence
>   Compromise of proprietary data.

THEFT is self-explanatory. It involves a deliberate unlawful act of taking the property of the company. When looking for controls in the computer-based part of the system, the only company asset exposed is information, and clearly the team must check for possibilities of the information itself being stolen. 'Stolen' in this case means removing the information so that it can no longer be used by the company; the definition includes acts of vandalism. A famous example of the theft of information occurred several years ago, when a computer operator erased all but one copy of a company's debtors file, then absconded with that copy held on magnetic tape, and attempted to extort a large sum of money from the company for the return of one of its most precious assets. It is important to bear in mind that loss of critical information could force many companies into bankruptcy.

However, the theft of the information itself is less likely than the deliberate manipulation of company data so that some other form of asset, such as stock or money, may be more easily stolen. Controls Analysis Teams will examine the areas of inventory and accounts with particular care.

LOSS OF ASSETS involves the accidental mis-appropriation of some of the company's assets, or the forfeiture of those assets as a result of some harm caused inadvertently by the company. In the broader business environment, this could refer to the misplacement of materials in some remote part of the warehouse through poor procedure control, it could involve the loss of money as a result of delivery delays, and it could also mean the payment of damages to a client who has been adversely affected by the company's product; (the major drug companies provide many such examples).

Loss of assets occur usually because of some loophole in the company's procedures, and obviously the computer-based part of the system will contain many such procedures. The whole subject of the accuracy of the computer programs must be considered, and the team must take into account the dangers of the computer system not functioning correctly (or even not functioning at all!). This is a very important aspect of the controls analysis study.

MIS-INFORMED DECISIONS are likely to be the cause of a number of problems within the system. Such decisions can be the result of the use of incorrect information, insufficient information, and on rare occasions too much information. One example of this type of situation is where an order is sent out even though the customer has cancelled it, the cancellation message having been delayed in the wrong department. A similar example relates to the bulk purchase of materials by the company buyer, on the assumption that a high level of production is needed, when the buyer does not have access to the latest sales figures. This could result in massive over-stocking.

The team must examine every process in the computer system, looking critically at the implications of wrong decisions being made. As the decisions involved are programmed, the problems will tend to occur as the result of incorrect information. For example a computer system may set in motion the delivery of ten times as many items as have been ordered, simply because of a transcription error during the order input.

EXCESSIVE COST INCURRENCE involves the carrying out of company procedures in an unnecessarily expensive manner. The Controls Analysis team must examine the proposed design not only from the point of view of accuracy, but also of cost-effectiveness. The team must satisfy themselves that the approach taken in the design is the cheapest necessary to carry out the objectives set.

COMPROMISE OF PROPRIETARY DATA refers to the unauthorised access to company information by those who may benefit from knowledge of that information. It differs from the earlier definition of 'theft', in that the information, although accessed, is not lost to the company, only its confidentiality has been breached. The Controls Analysis team must examine the risk of access to certain company information by for example competitors, and take measures to prevent or minimise the danger. Similarly, confidential information about members of staff and clients must be protected, or the company may be sued under the provisions of the Data Protection Act.

2.1.2.2 Identifying and Classifying the Threat

Each exposure point is examined carefully as to whether any of the five types of threat could result, wholly or partially, from an exposure at this point.

It should be noted that the most effective approach is to consider the proper 'outputs' first, then move on to look at the 'files'. In this way, the exposures relating to official access will be examined first, leaving only those related to un-official or accidental access to be dealt with on the forms relating to the files. This should minimise the danger of duplication.

Having identified the different types of threat relating to each exposure point, the team must then decide how important each of the threats is in terms of potential danger to the company.

There are three basic classifications used to describe the importance of the threat; high, medium and low.

> HIGH means that the threat is so great that the company could be seriously harmed if the worst case situation occurred.

> MEDIUM means that the company might lose a substantial amount in the worst case situation, but could bear the loss without major discomfort.

> LOW means that the company would expect this kind of threat to happen, and while it is perhaps prepared to put some resources into preventing them from happening, it

recognises that occurrences of this type are inevitable, and to some extent must be lived with.

2.1.3 The Threat Situations

Having identified each potential threat related to an exposure point, the team will then attempt to ascertain the circumstances in which the threat can arise. Although the danger itself may only occur as a result of the information leaving the system, the problem causing the threat may actually have been created anywhere within the computer system. For example, a 'customer delivery' may consist of a much greater quantity of stock than the customer originally requested. The problem may eventually be tracked down to a transcription error occurring during the sales order process.

It is the task of the Controls team to trace back through the system, identifying specific situations where a threat might be realised; (clearly these are the most likely places to install effective controls). In carrying out this process, the team will make use of the Systems DFD models, and in some circumstances they may use the ELH diagrams for the affected data stores.

In assessing the potential risk to the company, it is necessary to examine each of these threat situations and consider the probability of the threat being realised. Again, there are three classifications used, High, Medium and Low.

> HIGH means that the situation is likely to occur on a regular basis, and relatively frequently.

> MEDIUM means that the situation is expected to occur, but infrequently, and in no regular pattern.

> LOW means that the event is unlikely to occur, but there is always a possibility.

2.1.4 Controls Design

Now that the team has identified the threats, and the situations in which they are likely to arise, it can begin to make decisions as to the actual controls to use for each situation. In doing so, the members of the team must bear in mind that there is a balance to be established between the risk of losses and the cost of controls. It is often almost impossible, and certainly extremely expensive, to incorporate measures which will guarantee a completely controlled system. In some of the parts of that system, it would be costing the organisation more to irradicate the weaknesses than to let them happen!

2.1.4.1 Costs and Probabilities

For each threat, the team now have two important pieces of information to help them decide how much effort and expense to put into controls. These are:

> The Classification of the Threat in terms of cost to the company; (high, medium or low).

> The Probability of the Threat being Realised; (again, high, medium or low).

There are of course nine possible combinations of the two values, ranging from 'High/High', where it is absolutely essential to impose controls irrespective of cost, to 'Low/Low', where even the most simple control measures may prove more expensive than they are worth. It is not sensible to lay down detailed rules or guidelines for each of the combinations in all circumstances; the definitions are so open to interpretation, and the estimates of control costs are bound to be a matter of judgement. However, the team may find it worthwhile to draw up a range of costs they are prepared to consider for each of the combinations, though, in the end, each case must be examined on its merits.

2.1.4.2 Preventive and Detective Measures

For each potential threat, a decision must be made whether to use controls to PREVENT it from happening, or to DETECT that it has happened, then attempt to rectify or minimise the damage. A simple example of a preventive measure is the use of a password system to stop unauthorised access, and an example of a detective measure is the regular stock-count to check that the inventory figures match the warehouse contents.

Both types of control have their advantages and disadvantages, though one is tempted to assume that 'prevention is better than cure'. For example, the careful detailed examination of a very large number of transactions in order to prevent a rare defective case from entering the system may prove expensive, and cause some delay in the processing of the valid transactions, whereas, after a limited amount of processing, the small number of defective cases may become strikingly obvious and easily rectified.

CONTROLS ANALYSIS FORM				SYSTEM NAME	
				Materials Proc.	

OUTPUT / ~~FILE~~ NAME	Purchase Order

PURPOSE

   To order materials from a supplier to replenish stocks

CONTENT/DESCRIPTION

   Order No., Name & Address of Supplier.
   For each item type ordered;
   Supplier Item No., Our Item No., Quantity Required, Units

THREAT	DANGER H/M/L	SITUATION	DFD REF	PROB. H/M/L	CONTROLS
1. THEFT Items ordered then stolen on delivery	M	Purchase Order dialogue access by unauthorised person	3.5	M	Password system for buyers, unique value for each
		Similar access to Delivery Recording system	3.6	M	Password system for access to Delivery file, to control adjust- ments (no access for buyers)
2. MISINFORMED DECISION Incorrect stock quantities are re-ordered	H	Incorrect setting of Re-order Qty value in the 'Adjust re-order value' process	4.2	H	New process to create a mgmt report giving details of stock- outs & over- orders

Figure 10.20 A Completed Controls Analysis Form

On the other hand, in some circumstances detection can only occur when it is already too late to correct the situation. Here, preventive measures are essential. The controls team must be particularly careful in the use of detection controls in situations where fraud or theft could occur; many of the most effective frauds have been perpetrated by making use of the time-lag between the offence and the supposed detection.

Again, each threat situation must be examined, and the pros and cons of particular preventive and detective solutions considered.

2.1.4.3   Controls Effectiveness

As part of any control installed in the system, there should be a method of testing its effectiveness. All too often, expensive controls are implemented to prevent a particular danger to the system, and in reality the danger never materialises. If after a certain time the use of the control was reviewed and the threat probability revised, it might be possible to use a cheaper alternative.

A very common example of this can occur when a password system is implemented. Its purpose is to prevent unauthorised access to the system, but how can one tell whether or not it is effective? A well-thought-out password system should have a means of recording all the access attempts that are denied, and it should regularly report on such accesses to management. In certain types of system, the identity of the password holder should be associated with the transactions carried out on their authority.

2.2   Summary

When the team have completed their study, they will have produced a full set of Controls Analysis Forms (see example in Figure 10.20). Details of the controls to be included in the system are given to the analysts concerned, who will incorporate them into the appropriate computer processes, and then submit these revised versions as CONTROLS PROTOTYPES to the users.

Ideally, both the analyst and the user will have been involved with the controls team in the related discussions. As a result, the type of controls decided upon can be implemented and accepted with minimum communication difficulties. This is not always possible however, because a major element in the security of a control system may be the fact that as few people as possible know the details of its operation. It is of course important that staff believe that effective controls are in place, but if they knew how those controls worked, they might be able to devise a method of by-passing them!

3   ASSEMBLING THE COMPUTER SYSTEM

In software terms, a computer system is made up of a series of major suites or sub-systems, each of which consists of a number of computer programs. The programs themselves are groups of program modules, each module having been designed to carry out a specific function within the program. Up to this stage in the systems design process we have concerned ourselves only with these functional modules, shown as processes on the Systems DFD, and with the related non-computer processes with which they are incorporated as prototypes. Now we must consider how these prototyped modules are to be combined into programs, and how the programs will be best organised into sub-systems, etc.

The traditional approach to this has been to carry out the design of the computer system structure at an early stage of planning, before the individual modules had even been specified. The reason for this was that it allowed the proposed routines, programs and sub-systems to be used as units for estimation and control for the purposes of project management. Each module then became a 'deliverable', which would pass through the stages of specification, writing, testing, link-testing with other program modules, and final acceptance by the design team.

Such an approach is based on the assumption that the system is to be 'pre-specified'. When an evolutionary development method is adopted, decisions on the grouping of modules into programs and sub-systems can be left to a much later stage, when more information on the physical requirements, limitations and preferences of the company is available. In other words, the computer system structure can be better 'tailored' to the organisation's needs.

The assembly of the already prototyped modules into programs provides further opportunities for prototyping, this time at a higher level. Usually, in addition to the interactive functional modules so far discussed, the newly assembled program requires an extra module to handle aspects like the user's selection of options, the sequencing of functions, and the control of user access. The dialogue necessary for such a 'link' module clearly lends itself to the prototyping approach already outlined. The programs issued to the user for this purpose are referred to in this methodology as COMBINATION PROTOTYPES.

Also at this stage, it is quite common for the database accessed by the various modules to be altered, This may involve combining some of the previously independent file structures, which referred only to specific modules, into an integrated database suitable for the whole program. This may or may not be the optimised version of the complete database design for the whole system (as discussed in chapter 12). It does mean however that some of the file accesses already tested in earlier prototype versions may need to be re-checked.

The problems of assembling the computer system can be examined under three headings:

1. The Criteria used for grouping modules into programs and sub-systems

2. The Design approach adopted, and the extra concepts employed

3. The Modelling tools and techniques used within the methodology to assist and record the design.

All three are supported by examples from the Gentry case study.

3.1 Criteria for Assembling Computer System

It must be emphasised that the analyst's pragmatic judgement, combined with the user's recognition of the practicality of the solution are the main arbiters of whether the proposed computer system structure is best for the company. There are however three categories of criteria which can be used as guidelines to assist the analyst in putting forward options. These categories suggest that components should be assembled

> by Entity Group
> by Event
> by Expediency.

3.1.1 Assembly by Entity Group

The top-level components in any modern computer system are normally referred to as 'suites' or 'sub-systems', and each is usually based around one of the major 'Entity Groups' of the system. An entity group (also known as a 'Data Group') is a combination of all the entity types related to a particularly important form of asset or transaction being dealt with in the system. Earlier, in chapter 4 (which dealt with the construction of the Business Function Diagram), it was suggested that any major business function would do one of the following three things:

Provide a Product   (eg. Assemble a Component, or Manufacture a Unit)

Provide a Service   (eg. Repair Goods, Sell Materials, or Purchase Supplies)

Manage a Resource (eg. Manage Accounts, Maintain Stock, or Administer Personnel).

The analyst should identify the products, services or resources being dealt with by the computer system, and consider whether a sub-system should be built around each. There are clear advantages in having each sub-system dedicated to one important function of the company.

For example, the computer system being proposed in the Gentry case study involves three high-level functions, the Sale of Goods, the Purchase of Materials, and the Maintenance of Stock. The first two of these involve provision of services (Sales and Purchasing) while the third represents the management of a resource (Stock).

The next stage is to relate all of the major entity types which represent assets or transactions to the most appropriate of these candidate sub-systems (the resultant combinations being referred to as entity groups). Figure 10.21 illustrates a likely entity group split for the Gentry project.

ENTITY GROUP	ENTITY TYPE (FILE)
SALES	Sales Order Customer Sales Delivery Delivery Schedule
PURCHASING	Purchase Order Supplier Contract Purchase Delivery
STOCK	Goods Materials Production Order

Figure 10.21 Entity Groups for the Gentry System

It is not always a clear and simple decision as to which entity group a particular entity type should belong. There may be some question for example whether 'Sales Delivery' should not be considered as part of the 'Stock' entity group; after all, the work relating to it takes place in the warehouse, or is carried out by warehouse staff. The decision to include delivery as part of Sales may be based on the desire to make the Sales entity group (and sub-system) as integral as possible, avoiding an unnecessary split in responsibility between the customer's request for stock and the satisfaction of that request.

Using the same reasoning, the 'Production Order' entity type might be more appropriately placed within a Production entity group. However, as the production process is outside the scope of this particular system, the Stock entity group is left as the strongest candidate.

ENTITY GROUP	ENTITY TYPE (FILE)	RELATED MODULES	REF
SALES	Sales Order	Take Order Validate Order Item Handle Enquires Delete Order	2.1 2.2 2.4 3.1
	Customer	Validate Customer Adjust Customer Credit Handle Enquires Delete Customer	2.3 5.1 2.4 5.2
	Sales Delivery	Initiate Order Assembly Record Order Assembly Record Order Packing Record Delivery	4.1 4.2 4.3 4.4
	Delivery Schedule	Schedule Deliveries	4.5

Figure 10.22 Candidate Modules for the Sales Sub-system

Having arranged the main entity types of the system into their most appropriate entity groups, the designer can then list all the functional modules (or prototypes) which handle the creation, update and deletion for each entity type in the group. These are all to be considered as candidate components in the proposed suite or sub-system. Figure 10.22 illustrates the modules identified as related to the Gentry 'Sales' entity group.

Again it should be stressed that some modules may be considered to be equally appropriate to more than one proposed entity group. In such a case, an arbitrary decision may be made, and that decision may later be changed in the light of further information (perhaps resulting from experience with a Combination Prototype).

3.1.2 Assembly by Event

The second criteria to be considered is the relationship between the modules in terms of time. If different modules are 'triggered' by the same external event, or if the calling of one module is to be followed immediately by the calling of another, then there is a strong case for combining them in one program.

It is not uncommon for a particular event to set in motion activities relating to a number of different functions or sub-functions. A good example of such an event is the placing of a sales order by a customer. The main entity type concerned is obviously 'Sales Order', but if this is the first order placed by a particular customer, then it is necessary at that time to set up a new entry in the 'Customer' entity type (or file). Similarly, having received and accepted the order, the most efficient approach may be to trigger the assembly of the order immediately by producing a picking list on a remote printer in the warehouse. So, the arrival of the sales order triggers one major customer-related activity, and a delivery-related activity, as well as the expected sales order acceptance. The modules relating to these different activities will be combined in a single program (as shown in Figure 10.23).

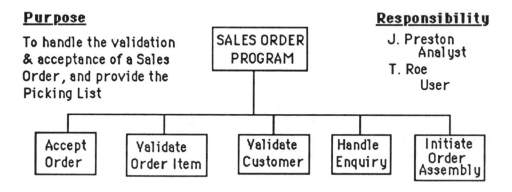

Figure 10.23 Gentry Sales Order Program Structure

It should be noted that the timing of many of these activities depends on how the organisation chooses to conduct its business. For example, if it is judged to be better business practice to group the sales orders into delivery schedules, and then to assemble all orders for a particular schedule together, it can be seen that the production of a picking list could not be included in the order acceptance program. It must be done as a batch process, as part of the delivery scheduling run.

3.1.3 Assembly based on Expediency

There are a number of other criteria that might be used to help decide on the best grouping of computer system components, but just how applicable each of these is depends very much on the individual circumstances of the organisation (for example, on its business structure, on the existing or proposed hardware, on the staffing situation, on the location of offices, etc.).

There are three particularly common types of situation which can effect the grouping of modules and programs. These occur

> where batch and on-line modules occur in the same function area, or relate to the same data,

> where staff with different levels of responsibility are working on the same data,

> where individual staff have a combination of duties which cross functional boundaries.

3.1.3.1 Batch and On-line Modules

Having decided that a group of modules relate to the same entity group, and may together form a computer sub-system, it is necessary to separate those modules which are to be performed in batch mode from those which provide on-line access. An individual program can not sensibly consist of a mixture of on-line and batch routines.

When there is a predominance of batch routines, then these also need to be separated into programs. Batch sub-systems are traditionally organised into separate programs for

> input of the transactions
> processing of transactions against the master file
> output of reports.

The need for this type of batch sub-system may exist because of lack of hardware facilities, or for business organisation or staffing reasons. Such situations are becoming rarer, and would not normally be designed using an evolutionary prototyping method. However, most computer systems which are predominantly on-line are likely to have some batch-oriented components.

3.1.3.2 Separation of Responsibilities

It is important that any group of interactive modules put together into a program should all relate to the same level of staff responsibility. It may be acceptable for a senior member of staff to access modules normally operated by junior members, but those same junior members must not be placed in a position whereby they may be able to access routines for which they have no authority. It can be argued that a password system inside the program might prevent this unauthorised access, but the very inclusion of these modules in the same program as is being used by the junior staff is an unnecessary risk.

3.1.3.3 Combination of Duties

In many on-line systems, most of the the user-operators are 'locked' into only a small part of the whole system. When the user's terminal is switched on, the first screen of a program appears, and the design is such that it is impossible for a user to escape from the program into other programs of the system. This means that the design of each of these 'transaction-processing programs' must be based on the duties and responsibilities of the particular user-operator. This 'custom-building' of on-line programs to suit specific jobs or even individuals within the company is a major advantage of the evolutionary development approach.

3.2 The Design Approach

So, the purpose of the sub-step is to assemble the computer modules and prototypes into programs and sub-systems, and the criteria to be used have already been discussed. The approach suggested for applying these criteria is to start with a rough grouping of the modules for each of the proposed entity group/sub-systems (as shown in Figures 10.21 and 10.22).

From that point onwards, it is usually better to work in a 'bottom-up' fashion, by deciding which modules should be grouped together to make a particular program; (the 'event' and 'expediency' criteria should provide the basis for the decisions). Then the proposed programs can be given suitable names, and the relationships between the programs can be considered. For example, in a proposed batch sub-system, it may be necessary to introduce special sort procedures between the different programs.

It should be remembered that throughout this step, the designer can always reconsider the grouping proposals, and re-organise the system into a different structure. There is one particular point at which this is likely to take place; it occurs when the designer has made an initial allocation of modules, then goes back to see if any modules from the Systems DFD have been overlooked. The requirement to find an appropriate position for a previously forgotten module surprisingly often causes a major re-examination of proposals.

The design organisation for the computer system is normally hierarchical; modules are grouped into programs, programs into sub-systems, and sub-systems into suites. The use of the term 'suite' is common within the industry to represent an extra level of grouping when the overall system is very large. Often the concepts of suite and sub-system are used to help divide up the computer system when allocating responsibilities for maintenance and support during 'live' operation.

There are two different types of module to be included in the programs being designed:

> FUNCTIONAL MODULES, where the processing involved is directed towards the achievement of some organisational goal. All of the processes shown in the Systems DFD fit into this category.

> LINK MODULES, which are non-functional, in the sense that they are there simply to provide the interface between the functional modules, and to control access to these functions by the user.

Much has already been said about the design of functional modules within the methodology. The only important point to bring out at this stage is that although most of these modules are unique, and relate to one specific situation, some modules will have a more general purpose, and may be included in many different programs and systems. Such modules are known as **sub-routines**, and are treated differently from the program-specific modules. From the early stages of the design, the analysts should be examining the proposed system for opportunities, either to use existing sub-routines, or to build new ones.

Link modules, on the other hand, will normally only be introduced during the 'Assembly of the Computer System' step, when the individual functional parts of the system need to be combined into practical programs. Often the designer must create a dialogue to allow the user to select the particular functional activity to carry out, or must set up a password dialogue to control user access to certain functions within the program. Sometimes a sequence of functional modules must be obeyed, and the control of this again should be programmed into an appropriate link module.

Link modules are important, in that they provide a place in which to code the 'administration' aspects of the program, and they therefore prevent the dilution of the functionality of the real process/modules in the program.

3.3 The Modelling Tools and Techniques

As with earlier stages in the development process, the analyst is assisted in the design process by the use of structured modelling techniques. The two types of model used are known as the COMPUTER SYSTEM DIAGRAM and the COMPUTER DATA FLOW DIAGRAM, and the ideas and symbols used in both will be very familiar. The techniques are in fact simply physical versions of the BFD and DFD discussed in such detail during the description of the Business Analysis stage.

3.3.1 The Computer System Diagram

This is a very simple form of model, the purpose of which is to illustrate the structure of the computer system, indicating which programs and modules are in which suites and sub-systems. It is a hierarchical decomposition diagram, and uses the same symbols and standards as are used for the 'Business Function Diagram' (described in detail in chapter 4).

There are normally two levels to the Computer System Diagram for a particular system. The top level is a single page showing all components down to 'program' unit (as illustrated in Figure 10.24), and the second level consists of a separate page for each of the programs, showing the modules of which it consists (as in Figure 10.25). Where the overall system is very large, it may be necessary to introduce an intermediate level, where the contents of each sub-system are illustrated on separate pages.

Each page of the diagram should contain a 'Statement of Purpose', describing in a single sentence or phrase the role of the top-level unit on that page. There should also be a reference to the section, team or person responsible for the development of the particular part of the system being described. At the top level this may be the name of a project team (with the name of the project manager quoted), whereas at the lower level it may be the names of the analysts and users involved in the prototyping development. Obviously this is not an essential element of the methodology standards, but there are benefits in attaching responsibility for completed work and recording it in the documentation.

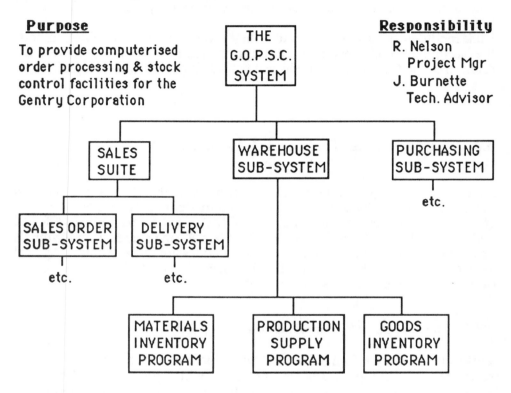

Figure 10.24 An Extract from a Top Level Computer System Diagram

The conventions used for naming the suites, sub-systems and programs are very flexible, though large organisations may find it necessary to impose stricter ones. Here it is suggested that a short, unique, descriptive phrase, followed by the word 'Program' or 'Sub-system' should suffice.

Below program level, it is strongly advised that the functional modules retain the names given to them at the time they were identified. These names enable analysts to relate the program modules to the original 'logical' view of the business. and this could be important when the system is later to be amended.

The same applies to any numbering system used for computer system components. It may be sensible to give each program a unique number, and to somehow indicate within that number the particular system and sub-system to which it belongs. However, the functional modules again may benefit from retaining the 'logical' number allocated during the business analysis stage.

It should be noted that 'Link' modules are not shown explicitly on the Computer System Diagram. It may be assumed that there will be one link module per program, controlling access to the functional modules; the link module can then be said to be represented by the higher-level box containing the program name. Where there is a need for a further link module, perhaps controlling a group of lower-level modules, then a box describing this group as a 'sub-program' might be used in the hierarchy, and this could be said to represent the new link module.

The Computer System Diagram is a very important model. It is included in the documentation of the system, and constitutes the highest-level description of the full computer system. As such, it will prove essential to analysts who are required to amend the system during its operational life.

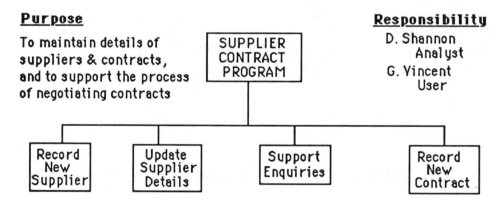

Figure 10.25 Computer System Diagram for Supplier Contract Program

3.3.2 The Computer Data Flow Diagram

This is the main type of model for describing the structure of the computer system. It is based closely on the earlier types of Data Flow Diagram modelling techniques, and all of the symbols have the same meanings as were described in detail in chapter 5. There are of course some small differences of interpretation, and these will be covered here.

The Computer DFD bears the same relationship to the Computer System Diagram as the Business DFD does to the Business Function Diagram. In other words, the two models must be completely consistent, and should be used to cross-check and correct each other. In particular, an analyst may build a CSD as a proposal for grouping modules into a program, and the process of building the Computer DFD would help either to confirm or revise that proposal. Figures 10.25 and 10.26 illustrate this relationship.

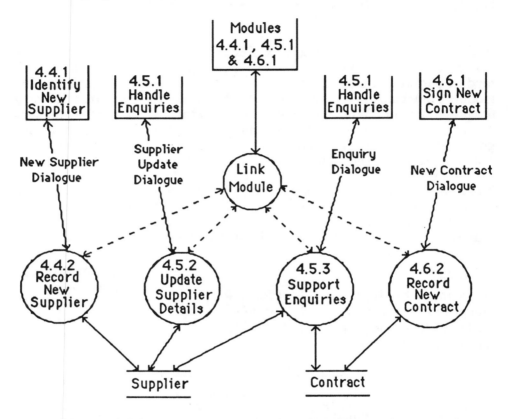

Figure 10.26 Computer DFD for Supplier Contract Program

3.3.2.1 The Computer/User Interface

Perhaps the most important difference between the Computer DFD and the earlier Systems DFD is that the non-computer processes, previously shown as circles on the 'user' side of the dotted line, become 'Internal Agents' on the new model. Figure 10.27 illustrates this.

This change reflects the fact that we are not interested in these processes as part of the 'computer system', but need to show them as the sources and destinations of our inputs and outputs. At the lowest level of the Computer DFD model, these internal agents are given the names of the user processes as shown on the Systems DFD, but are also marked with the name of the new section or department in which they occur within the company. At the higher levels of the model, only the section or department names may need to be shown (otherwise there is a possible danger of the 'sunburst' effect at these levels).

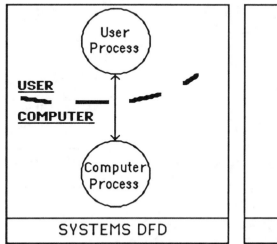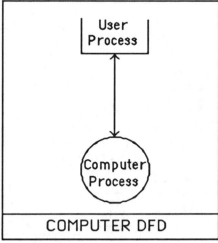

Figure 10.27 User Processes in Systems and Computer DFDs

3.3.2.2 The Link Module

It is essential to include the Link module(s) in the Computer DFD, but this inclusion can make the model look more complicated than it really is. Obviously the link module must be accessed by the internal agents (user processes), in order to gain access to the appropriate functional modules. However, once this access has been established, the dialogue is conducted directly between the internal agent and the functional module. This means that two flows are needed to model each user-computer dialogue. Figure 10.28 demonstrates this in the model for the proposed Sales Order Program.

3.3.2.3 The Flow of Control

A distinction is made in the Computer DFD between the *flow of data*, indicated in the traditional manner by a line, and the *flow of control*, shown by a dotted line. When a module calls another module, it passes control to

243

it. When the called module is completed, control will be returned to the caller. This is also illustrated in the model in Figure 10.28.

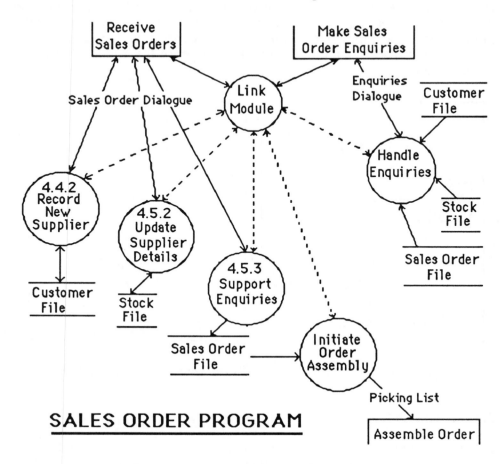

Figure 10.28 Computer DFD for Sales Order Program

3.3.2.4 Files shown on the Model

The files shown on the diagram have the same appearance as those on the Systems DFD, but in this model they reflect the 'physical' files that are to be used in the system. This means that the files shown are either the 'dummy' files from a limited prototype, or the final optimised database structure as is discussed in chapter 12. Obviously, in the final completed version of the documentation, the model must reflect the form of the files as implemented in the live system. However, at this stage in the development those final decisions may not yet have been made.

One important point relating to this is that the format and structure of these files will depend very much on the particular physical file handler or database package to be used. There is in fact a very wide range of possible forms of file structure, and this means that this aspect of the model's appearance must be left open to flexible interpretation.

3.3.2.5 The Layout of the Model

There is of course a great deal of freedom in the designer's use of layout when building a computer DFD. However, the generally recommended guideline is to position the 'internal agents' at the top of the diagram, then the link and functional modules (processes) in the centre. The files can then be placed at the bottom of the diagram. This approach has been found to minimise the problems of crossed 'flows'. A very clear example of this is shown in Figure 10.28.

3.4   Summary

This stage of the development process involves the combining of previously separate prototypes into computer programs, sub-systems and suites, and the grouping criteria, the design approach, and the modelling techniques to be used, have all been discussed here in some detail.

The last point to stress concerning this stage is that the prototyping of these combined programs is one of the most important activities in the whole design process. By this stage, all of the early weaknesses of the design should have been cleared, and this particular version of the prototype should have almost the exact appearance of the proposed live system. (The later 'Optimised' version is likely to be concerned only with efficiency improvements in the database and code structure, so the changes involved will be for the most part transparent to the user).

This means that the user and analyst together must work on the detailed system testing of the programs, and user management can start to finalise plans for the acceptance of the system. Operating instructions, user manuals and training programmes can also be taken to an advanced stage of development, again with the user making the major contribution. In fact, most of the tasks that traditionally have been necessarily held back until the Implementation stage of the Systems Development Life Cycle, can be performed at this much earlier stage when an Evolutionary Development approach has been taken.

11 ANALYSING DATA USAGE

Even before the detailed design of the computer processes has begun, work will have started on the problem of designing the most effective database structure for the proposed system. Clearly, the Entity Model and the Relational Model from the Business Analysis stage will be used as the basis for this design. However, the designers also need information on the overall usage of data within the system, and this means identifying in detail how data is to be used by each of the newly designed computer processes.

This work can begin as soon as the analysts have identified the computer system boundary, and separated out the computer processes or modules for prototyping; (see Figure 11.1). The prototyping itself may be done using simplified versions of the entity model, producing from them a number of 'dummy' database files. However, before the full physical database can be designed, it is necessary to check that the data model on which it is to be based is able to satisfy the detailed requirements of the computer system. This is done using an approach known as Logical Path Analysis.

Logical Path Analysis involves examining the entity model and plotting the method of access to data for each of the computer systems processes. This is done for two reasons:

> to check the validity of the entity model

> to provide detailed information on access requirements for
> the physical database design process.

All the main structured systems methodologies use different modelling techniques for this purpose. In Systemscraft, three major techniques are used, each one providing a slightly different angle on data usage within the system. Together all three provide the database designer with all the information needed to create and optimise the database. (Bear in mind that whenever the term 'database' is used here, it includes the possibility of the more traditional file approach of index-sequential and random organisations.)

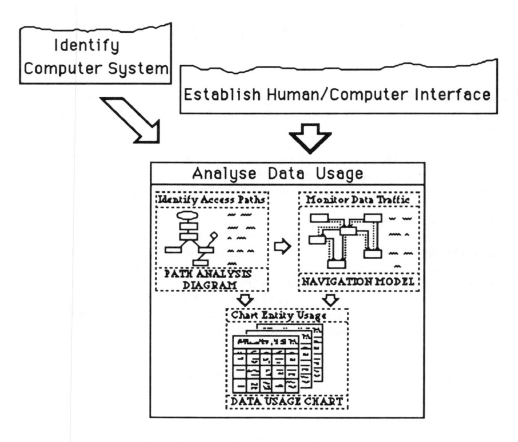

Figure 11.1 The 'Analyse Data Usage' Step in Systemscraft

The three techniques are;

1. **Path Analysis Diagrams** (also known as Path Analysis Maps). These show how information for the carrying out of a process is obtained by making use of the relationships between the various entity tables concerned. A piece of information (attribute) in one entity table can point to an entity in another table where more detailed information can be found.

2. The **Navigation Model.** This is an annotated version of the data model, showing the entry points at which the database will be accessed, and indicating which relationships

248

on the model are used in obtaining the necessary information to carry out all processes.

3. The **Data Usage Chart**. This contains details for each entity type in the model which is accessed. It includes information on the number of entities accessed, the frequency, the attributes made use of to effect the access, and a number of other aspects which will aid the designer in building a flexible and efficient database.

Each of these techniques is explained briefly, and a number of worked examples are given. It should be stressed that although the techniques of Navigation Modelling and Data Usage Charting are described separately, it is very likely that the analyst/designer will work on them simultaneously.

1   THE PATH ANALYSIS DIAGRAM

Normally at least one path analysis diagram is built for each computer interaction (or prototype) identified in the Systems Data Flow Diagram. It consists of three activities:

1. Identifying what information needs to be produced as output from the process being examined.

2. Identifying what information is provided as input to that process.

3. Tracing through the entity model, showing how the required information can be obtained from different entity types/tables (it is useful to think of each entity table as a logical file or sub-file).

It is possible to access specific information directly from a table by quoting the unique identifier (key) of the appropriate entity. Alternatively, if the unique identifier is not available, one can quote a non-key attribute, and the table can be searched for all entities with that particular attribute value.

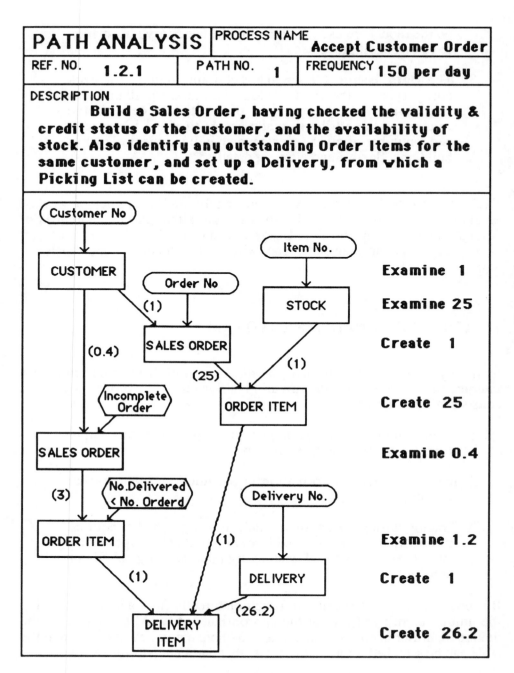

PATH ANALYSIS	PROCESS NAME Accept Customer Order	
REF. NO. **1.2.1**	PATH NO. **1**	FREQUENCY **150 per day**

DESCRIPTION

Build a Sales Order, having checked the validity & credit status of the customer, and the availability of stock. Also identify any outstanding Order Items for the same customer, and set up a Delivery, from which a Picking List can be created.

Customer No

Item No.

CUSTOMER — Examine 1

Order No

STOCK — Examine 25

(1)

SALES ORDER — Create 1

(0.4) (25) (1)

Incomplete Order

ORDER ITEM — Create 25

SALES ORDER — Examine 0.4

(3) No.Delivered < No. Orderd

Delivery No.

ORDER ITEM — Examine 1.2

(1) (1)

DELIVERY — Create 1

(26.2)

DELIVERY ITEM — Create 26.2

Figure 11.2 Example of a Path Analysis Diagram

250

The path analysis diagram plots the order of access of the different entity tables, showing the frequency (how often the transaction is carried out), and the nature of the various accesses (whether an entity is created, examined, updated or deleted). This aspect of the modelling technique makes it suitable for use as a kind of high-level 'flowchart' in the design of the computer processes, enabling designers to record their decisions about the sequence of data accesses, and about the proposed shape of the program/module.

1.1 Example of Path Analysis Diagram

Figure 11.2 gives an example of a path analysis diagram for a fairly complex computer transaction 'Accept Customer Order' (this would exist as a computer process on a Systems DFD).

It illustrates how the customer identifier is presented and the customer entity table is checked to ensure that it represents a valid customer, and that the customer credit status is acceptable. Other information from the customer entity may be extracted and used for further checking, or for inclusion in output documents (eg. a receipt or delivery note).

Each item of stock requested in the order is examined against the stock entity table, again for validation, but also to see if there is sufficient quantity of stock to satisfy the customer's requirement.

The new order entity itself can then be created, followed by entities for each of the lines in the order.

At this point, the process checks for any outstanding items from previous orders made by this customer. (These need to be associated with the new order for the purposes of picking and delivery.)

Having now identified all the items that require to be delivered, a delivery entity can be created, and the order items, both new and outstanding, can be used to set up delivery item entries.

1.2 Path Analysis Diagram Symbols

The path analysis diagram consists of a small set of symbols, several of which are very similar to the those used in earlier tools and models.

251

1.2.1 Unique Key Access Point

The first symbol is used to indicate an entry point into the diagram, and represents the piece of information in the possession of the process user, enabling the required entity to be accessed. Figure 11.3 illustrates the symbol used, and gives examples of possible values. The identifier quoted in this symbol MUST provide UNIQUE access to an individual entity in the table.

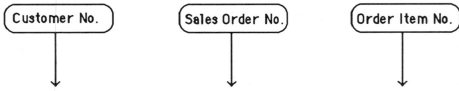

Figure 11.3 The Symbol for a Unique Key Access Point

In some circumstances where a new entity is added to a table, the identifier to be allocated is the next available number in a sequence. This new number must obviously be obtained from the computer system, so it is not really an 'access point'. However, as the creation of the entity does require the allocation of such a (unique) number, the Unique Access Point symbol is used to indicate this. (For example, in Figure 11.2, both 'Order Number' and 'Delivery Number' are allocated in this way.)

1.2.2 Non-unique Access Point

Often the user does not have a unique identifier with which to access an entity; for example, it may be that a number of entities from the same table are required. In this case a non-unique identifier is presented, and a different symbol is used to indicate it (see Figure 11.4). A non-unique identifier can simply be an actual value of a non-key attribute, or it can be a condition on that value (eg. less than 20, within range 10 to 50, etc.).

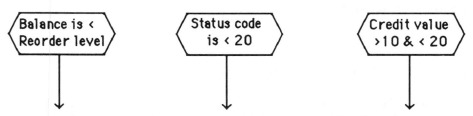

Figure 11.4 The Symbol for a Non-unique Access Point

A comparison between these two symbols is given in Figure 11.5, contrasting the situation where the user is able to quote the customer's number, with one where the customer's name (which may not be unique) is all that is available to the user. It is not uncommon to find that a particular entity access requires more than one access point box. This is perhaps not surprising when one realises that these boxes contain the **selection criteria** for one or more entity occurrences.

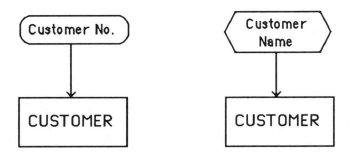

Figure 11.5 Comparison between Unique and Non-unique Accesses

1.2.3 Entity Types and Accesses

The entity tables that are addressed are symbolised by boxes, and next to each box is a text indication of the number of entities which are to be made use of in this one occurrence of the path transaction. This concept of indicating the number to be 'made use of' is important, and is not the same as the number to be 'physically accessed'. Sometimes it will appear to be necessary to examine every entity in a table in order to identify two or three occurrences that are required. Only the number of occurrences required are noted on the path analysis diagram.

The nature of the accesses are also noted. An access can be to

> Create
> Examine (no change)
> Update (change)
> Delete

an entity.

Figure 11.6 gives some examples of the use of the symbol.

253

Figure 11.6 The Entity Type Symbol plus Notation

1.2.4 Information Lines and Cardinality

The third way in which an access is shown on a path analysis diagram consists simply of a flow line between two entity table boxes. This indicates that the information being used to find the occurrence(s) in the second table has been obtained from an entity in the first. These are the important accesses for the purposes of checking the validity of the data model; they indicate that a **relationship** within the data model is being made use of.

Figure 11.7 gives two examples. The first shows how a customer entity has been identified, and the customer number is used to examine the Order entity table to find the appropriate order. On the other hand, the second example shows how an order entity has been accessed (directly by means of order number), and the Customer number present in the order entity is then used to identify the appropriate customer entity for further details.

The numbers in brackets indicate the average number of entities to be found in the lower table for each entity accessed in the higher table. (This aspect of the relationship between two entity types is sometimes called its 'cardinality'.) In Figure 11.7, the value in the first example is shown as (3), meaning that for any customer entity obtained there is likely to be on average three order entities. However, in the second example, for each order there can only be one related customer entity.

Sometimes the number to be quoted in brackets can be less than one, indicating that there are often occasions when there is no lower table entity to be accessed using the higher table entity. This occurs most commonly when there is a combination of selection factors applying to the table, one or more of which are non-unique. An good example of this occurs in Figure 11.2, where there are two separate lines shown from the Customer entity table to the Order table. One of these represents the new order being created, so the number used is (1). The second line however represents the access of any outstanding orders for that customer, and as the number (0.4) illustrates, most customers have no orders outstanding.

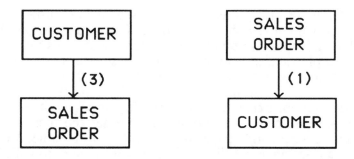

Figure 11.7 Examples of Lines connecting Entity Tables

1.2.5 Sequence

It must be emphasised that the PAD model also shows the order in which the entity tables are accessed, by placing one entity type below another in the diagram if it is to be accessed after it.

For example, Figure 11.2, 'Accept Order', shows that the customer details are checked first, then the individual items are checked against the stock file. When all the items have been checked, the operator can confirm the acceptance of the order, so the order, with its individual items, can then be added to the appropriate entity tables. After that, old orders with outstanding items can be searched for, to be added to the picking list. Then the delivery and delivery line entities can be set up.

If two entity tables are accessed simultaneously, they can be placed next to each other on the diagram.

This sequential aspect of the model is particularly useful in the design of the computer process, and the Path Analysis Diagrams may actually be built during the earlier 'Establish Human/Computer Interface' stage of the systems development.

1.2.6 Summary of PAD Symbols

Figure 11.8 gives a graphical summary of the symbols used in the Path Analysis Diagram.

It should be noticed that there are no symbols for showing 'temporary' files. Such files are by nature physical, and this is a logical modelling technique. Once an entity has been accessed, it is assumed that all the information in that entity becomes available for the duration of the process being modelled. Though in practice some temporary files may prove to be necessary in a physical prototype, it can be assumed while building these models that there is no limit to the main memory available.

Similarly, no symbol exists to connect different pages of the same PAD model. The absense of such a symbol emphasises the fact that no model should extend over a single page; if there seems to be a need for a larger model, the transaction should be examined with a view to splitting it into into a number of smaller transactions.

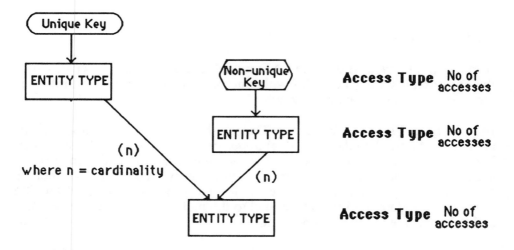

Figure 11.8 The Path Analysis Diagram Symbols

1.3 Building a Path Analysis Diagram

The process of constructing the path analysis diagram consists of examining the computer process specification (ie. the Structured English description of the DFD process), and the relevant part of the Data Model. The process description should indicate what information is GIVEN at the start of the transaction, and what information has to be provided at the completion of the transaction. The designer must plot the path from the GIVEN information to the REQUIRED information via the relationships and entity types of the data model.

A line on the path analysis diagram must have an equivalent line in the data model, though it does not matter whether the line on the PAD is going in the 'one-to-many' or 'many-to-one' direction. Figure 11.7 illustrates this.

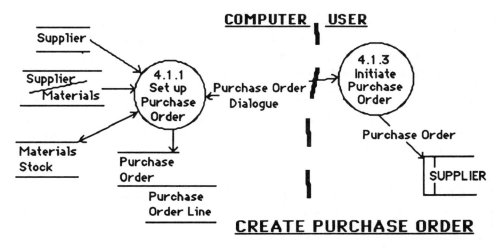

Figure 11.9 The Systems DFD for 'Create Purchase Order'

Building a PAD can best be explained by means of an example, and the example chosen is that of the 'Create Purchase Order' process for a simple stock control system. The information required at the start of the modelling activity is present in the Systems Data Flow Diagram (Figure 11.9 shows the relevant excerpt), the appropriate part of the Data Model (Figure 11.10), and the Process Description (Figure 11.11).

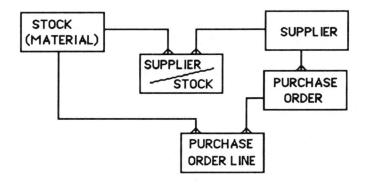

Figure 11.10 The Data Model for 'Create Purchase Order'

Figure 11.12 illustrates the path analysis diagram which might be built using this information. The process description indicates the need to find the stock items which are below re-order level. So we begin by examining the Stock entity table, selecting cases where the stock balance attribute value is less than the re-order level value.

Having identified the appropriate stock item types, it is necessary to find which suppliers are able to supply them. The only way to do this is to access the Supplier/Stock entity table, then to select from the candidates one supplier who is able to supply the most required items.

Having chosen a supplier, we can then build a purchase order , and create purchase order items for all the re-order cases for which that supplier caters. It is also necessary to update the Stock entity to indicate that an order has now been made for it.

PROCESS DESCRIPTION   ~~Logical~~ Physical

Process Name   Set up Purchase Order | **Ref.** 4.1.1

1. Examine Stock table to identify stock entities with quantities below re-order level.

2. Identify Supplier able to provide the greatest number of re-order items.

3. Create a Purchase Order for that Supplier.

4. For each re-order item for that supplier,

   4.1 Create a Purchase Order Line

   4.2 Update Stock entity with the new order number.

Figure 11.11 Process Description for 'Create Purchase Order'

In the earlier example of a Path Analysis Diagram (Figure 11.2) a formal document layout is made use of, and all the necessary references are completed. There is strong argument for using such formality; the model which has originally been created by the analyst who is building an initial prototype, may be passed to others who are working on the database design. However, as always, when developing a small system, a much less formal approach can be taken.

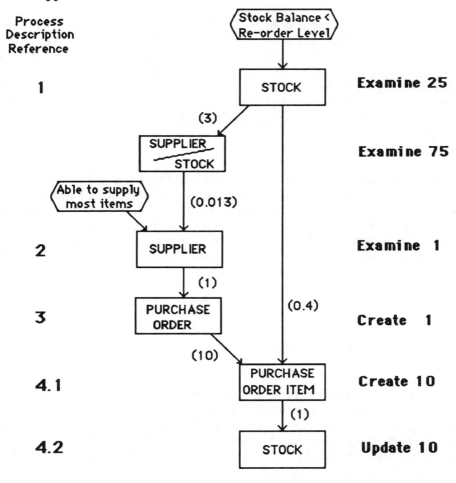

Figure 11.12 Path Analysis Diagram for 'Create Purchase Order'

1.4 Conclusions

Although the two examples shown so far in this chapter involve fairly complex logical access paths, it is quite common for processes which update files to have relatively simple paths, accessing only one or two tables. On the other hand however, some enquiry processes making access to a corporate database can have a number of very detailed and intricate paths.

It has already been mentioned that the PAD technique can in some circumstances be used as a program design aid, along the lines of a high-level program flowchart. It has the big advantage over the traditional flowchart in that it is based on the structure of the data to be used in the program module to which it refers. Some fourth generation environments (for example Focus) may be particularly suited to this kind of diagram, and in those circumstances it may be built at an earlier stage, as part of the prototype planning.

In most small systems developments, the building of a full set of Logical Path Analysis diagrams may not be necessary, and may in fact be a waste of resources. On the other hand, there are circumstances where detailed accurate modelling of the paths is absolutely essential. It is left to the analyst to make this judgement about each individual project. An understanding of the material in the rest of this chapter should hopefully provide guidelines on which to base the decision.

2 THE NAVIGATION MODEL

The Path Analysis diagram provides a picture of data usage for each process in the proposed system, but there is also the need for an overall picture of data usage throughout the system, shown perhaps in one single page diagram. The Navigation Model is designed to provide this. It pulls together information about all of the individual access paths that form the different processes, and imposes them on the Data Model (provided from the analysis stage).

This shows which relationships in the model are being made use of, and (sometimes more importantly) which are not. It also gives an indication of the number of paths that use each relationship, and their identities, making it easy for the designer to guage how heavily the potential files will be utilised. In fact, this model, along with the the logical path analysis diagrams and the data usage charts (to be described shortly), should provide all the information that a database designer needs to build and optimise the database structure.

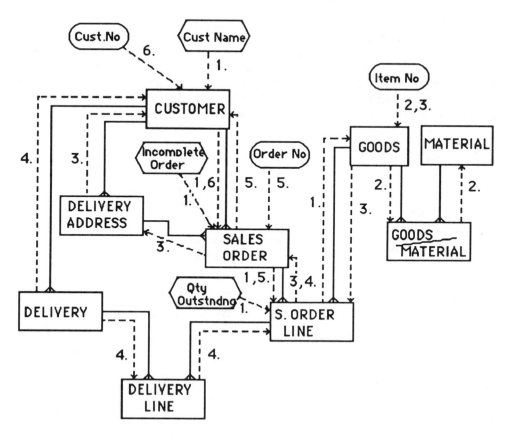

Figure 11.13 An Example of a Navigation Model

2.1 Definition of Navigation Model

The Navigation Model is simply the Data Model of the system, with the individual access paths superimposed upon it. Figure 11.13 gives an example of such a model:

1. The data model components are exactly as they were in the analysis stage.

2. The access paths are traced on the model, using a dotted line with an arrow to indicate the path. The numbers beside the dotted lines identify the particular paths involved. The access entry points are also recorded, using the same symbols as in the path analysis diagram.

The building of the navigation model also acts as a check on the validity of the construction of the individual path analysis diagrams. It is not unusual to find, when imposing a particular path onto the model, that an error has been made, and a false relationship path has been built into the diagram.

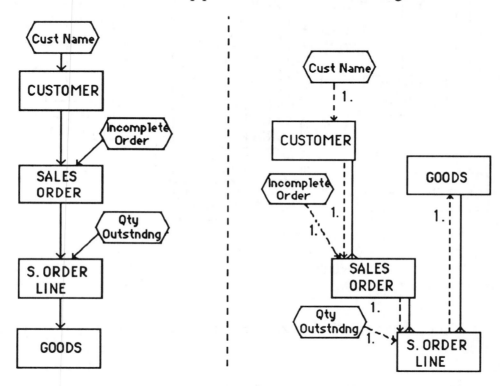

Figure 11.14 An Enquiry PAD, and its Incorporation within the Navigation Model.

2.2 Building the Navigation Model

This is a very straightforward process. The data model is used as the starting point, and each path analysis diagram in turn is added to it. This is

262

best illustrated by example. Figure 11.14 gives an example of a simple PAD for an enquiry, 'Find whether there are goods in stock for any of the outstanding orders for this particular customer'. It also shows how part of the navigation model for the system would look when this enquiry has been incorporated. The modeller starts at the top of the PAD, mapping in the access points, and gradually works down through it, matching the entity types with those on the model. Note that it is important to record the direction of the path, and that it may be from the one to the many (as between 'Customer' and 'Sales Order'), or from the many to the one (as between 'Sales Order Line' and 'Goods').

Once the navigation model is complete, it can be examined for any gaps or inconsistencies. The fact that a particular relationship has not been made use of can mean one of two things:

1. either the relationship is not very important, and may be ignored in the building of the physical database structure,

2. or some important processes have been overlooked, and the full set of path analysis diagrams have not been included.

2.3   Conclusions on Navigation Models

Navigation models follow on naturally from path analysis diagrams, and summarise them effectively, giving the designer an overview of the data design problem. However, there are some situations and circumstances where the use of one or both of these tools must be modified.

1. When the system being developed is small, and the file performance is known not to be critical, the designer may ignore the use of the path analysis diagram, and simply trace each process path against the data model. In a very simple non-critical system, it may not even be necessary to build the navigation model.

2. When the system being developed is large and/or complex, individual path analysis diagrams will be essential. However, when these paths are all mapped onto the same data model, the resulting diagram can be so detailed as to be

unreadable! In these circumstances, it is suggested that higher-level processes should be used as a way of partitioning the navigation model, and several different navigation models can be produced, one for each sub-system in the development. When this happens, there is clearly an overlap between different models (ie. the same entity types will appear in them) so much more weight has to be given to the third of the data path analysis tools, the Data Usage Chart.

3. One other simplification can be applied to the navigation model, both when the system is very simple, and when it is very complex. Instead of noting every path making use of a relationship, a broader view can be taken, and each path between two entity types can be marked as 'heavy traffic', 'light traffic', or 'no traffic at all'. This can be indicated by different thicknesses of line, or by the use of colour code. In cases where the hardware and software performance is not absolutely critical to the success of the system, this simpler approach may provide all the information the designer needs.

3 THE DATA USAGE CHART

The third model produced in the investigation of the usage of data in the required system is the Data Usage Chart. Its purpose is to identify for each entity type (and therefore each 'logical' file) all of the accesses and updates which are to be made to it during the running of the system. This information is essential for the designer when making decisions on optimising the physical data structures for the new design. Figure 11.15 gives an example of a Data Usage Chart.

The Data Usage Chart is built using the Navigation model and the Path Analysis Diagrams. Each entity type in the Navigation model will have a number of accesses made to it, all shown by dotted lines or entry point symbols. These have been derived from the Path Analysis diagrams, and represent all the accesses made to that entity type within the system. Figure 11.16 shows an excerpt from the Navigation model shown earlier, and it illustrates how each of the 'path' accesses is recorded against the name of the entity type being accessed.

A factor to beware of is that often only the most important of the system's transactions are recorded in path analysis diagrams and navigation models. However, the Data Usage chart must contain all the data accesses, including the most trivial. Sometimes this means re-examining the computer process descriptions associated with the Systems DFD model.

DATA USAGE CHART

ENTITY TYPE	ATTRIBUTES	Order No., Order Date
SALES ORDER	Customer No., Delivery Address Code, Order Status	

Source Data for Access	Process Id.	Path No.	Frequency	Entities Accessed	Total
Order No. (next available)	Take Order	4	150 per day	1	150 per day
Customer No. + Order Status value 'incomplete'	Take Order	4	150 per day	0.4	60 per day
	Handle Enquiries	1	20 per day	0.5	10 per day
Order No. + Order Date value 'within last month'	Handle Enquiries	3	25 per day	6.5	163 per day

Figure 11.15 Example of a Data Usage Chart

Figure 11.15 shows a Data Usage chart for the 'Sales Order' entity type. It lists the accesses to the entity type, showing what information is used in order to effect the access. For example, some accesses are made by quoting the order number of the entity required, others are made by quoting the customer number and a range of dates (enabling access to be made to all the orders for that customer within the dates specified).

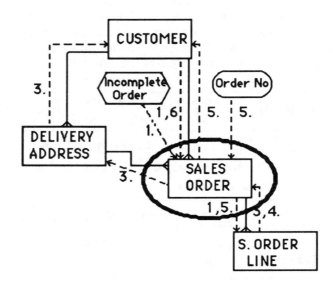

Figure 11.16   Excerpt from Navigation Model showing the Related Data
Usage Chart Entity

The first access shown on the chart indicates the use of the order number, but this is a slightly unusual form of access. The purpose of the access is to create a new order, so the order number it is to be given is one greater than the last number used. Somehow the information as to the highest number used so far must be held with the file.

Every access must exist in and be part of a particular computer process, and the name of the process is recorded in the chart. The Access Number, as used when marking the path of the access on the Navigation Model, must also be recorded, to help check for completeness.

The last three columns of the chart contain the figures which are to be used by the designer to help identify the most appropriate physical design option. The 'Frequency' column refers to the number of times the transaction takes place, and this may be in the form of the volume per hour, per day, per week, etc., whatever is the most relevant.

The 'Entities Accessed' column indicates the number of table entries actually required in the transaction. It should be noted that this does NOT refer to the number of entities which must be examined in order to find these required cases; that is a physical aspect of the eventual file design. Also there will be times when the transaction involves searching for an entity and not finding one. A good example of this is the second access to the sales order table shown in Figure 11.15. Here the process of taking orders includes the search for any earlier orders for the same customer which have not yet been fully satisfied. In many cases no such unfulfilled order will exist. The value quoted in this column is the average number, which in some cases (as in this) will be less than one.

The final 'Total' column simply records the two previous column values multiplied together. It does however show at a glance the number of entities made use of from this table in a given unit of time. Although the models being discussed here are logical not physical, these total figures do give us information which represents a kind of physical design constraint. For example, the Data Usage Chart in Figure 11.15 indicates that a Sales Order file must be designed so that 386 orders can be accessed from it each day!

4   CONCLUSION

The process of Analysing the Data Usage can start as early as the first computer system design proposals are put forward. It involves looking at each proposed computer process, and considering it as one or more transactions against the system's logical database. The information about the number and complexity of these transactions is built up for the whole system and finally, by means of the Navigational Model and the Data Usage Chart, it is 'inverted', to show for each entity type the accesses which will be required of it. This change of view of the data, from that used for an individual 'process', to all accesses to a logical 'file' (entity type), is vitally important for the process of physical database design.

On rare occasions it may be necessary to examine the usage of the most important entities in a proposed system, just to see whether such a system is feasible, or even possible! Where this is the case, the data usage analysis must be done as part of the feasibility study (and therefore must be carried out within the first few days of the project).

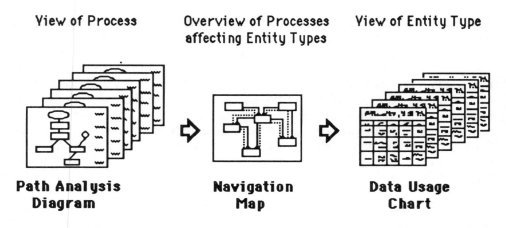

View of Process   Overview of Processes   View of Entity Type
affecting Entity Types

Path Analysis   **Navigation**   **Data Usage**
Diagram   **Map**   **Chart**

Figure 11.17 Relationship between the Data Usage Models

One last point of warning should be made concerning the tools for data usage analysis. Throughout the modelling process we have concentrated on AVERAGE numbers of accesses for each transaction. There are however situations where, although an average usage level would be acceptable, the incidence of peaks and troughs can lead to spells of much heavier-than-normal traffic which may cause unacceptable response times. Where such a danger exists, the collection of maximum/minimum as well as average figures may be necessary.

12 COMPLETING THE PHYSICAL DESIGN

This chapter examines the two final steps in the completion of the physical design of the computer system:

1. The design of the physical database

2. The completion of the design of the computer processes

It is in fact one of the shortest chapters in the book, the reason for this being that methodologies are designed to address the general rather than the particular, whereas the later stages of physical design must of necessity be dependent on the specific hardware/software environment available to the organisation. As a result, much of the physical design process is beyond the scope of a book which is primarily concerned with methodologies. Having said that, an attempt is made to show the limit of the general version of the methodology, and the ground is prepared for later discussion on the need for 'tailoring' of the methodology to suit each particular fourth generation environment.

Figure 12.1 is an excerpt from the System Design Road Map showing the two steps, and indicating how they relate to the earlier design processes.

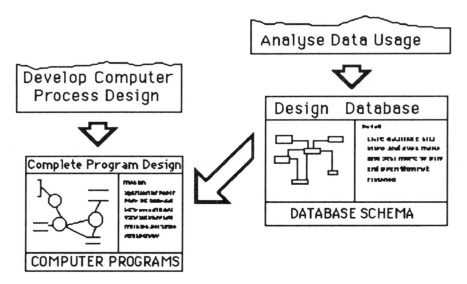

Figure 12.1 The Late Physical Design Steps

1 DESIGNING THE PHYSICAL DATABASE

Physical database design is the process whereby the file/database designer creates the data definitions for the proposed live system, and sets up the master file structures ready for implementation. This involves making use of the following information:

1. The user's performance constraints, in terms of hardware and software, response times, security and integrity controls, etc.

2. Details from the analysis of data usage carried out earlier; ie.

> the Data Model
> the Relational Model
> the Path Analysis Diagrams
> the Navigation Model
> the Data Usage Charts.

The designer will normally start by applying what are referred to as 'first-cut rules' to the logical data model, to convert it into a preliminary set of files in accordance with the particular file handling software being used by the organisation. From then on, the database structure will be 'massaged' and optimised until it satisfies the performance requirements of the system.

To begin with, this optimisation process is likely to be a paper exercise, whereby the data usage models are adjusted and re-calculated. However, in time-critical systems it is often necessary to build a 'performance prototype' of one or more of the most important files, to make sure that the required file constraints can be met.

Obviously the exact procedures adopted by the file/database designer are heavily dependent on the particular software to be used, and on the hardware that is available. As a result it is not possible to be specific about this process; all that can be provided are guidelines and examples.

1.1 Database Management Systems

There are three common types of structure for database management systems. These are:

1. Hierarchical (eg. IMS, FOCUS)

2. Network or Codasyl (eg. IDMS)

3. Relational (eg. ORACLE, DB2)

The techniques used to convert the data/relational model into a series of physical file specifications vary considerably depending upon the type of structure to which the proposed file handler adheres. For example, if the DBMS to be used is of a relational structure, then the process of conversion can be a simple one. In the best of situations, each of the entity types in the model will become a file (or sub-file) in its own right.

On the other hand, if a DBMS with a hierarchical or network structure is to be used, then the conversion procedure may be much more complex, though this is not necessarily always the case.

1.2   The Design Process

In this chapter, the example of a (hypothetical) hierarchical DBMS is used, and the various design techniques and stages are illustrated. The data model to be used is one with which we should now be very familiar, and it is shown in Figure 12.2.

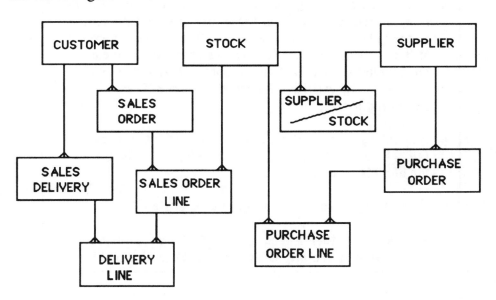

Figure 12.2 The Data Model starting point for
Physical Database Design

1.2.1 The First Cut Version

When the conversion process from logical model to physical database is a complex one, then it is usually carried out in two stages. The first of these stages involves the application of a set of standard guidelines known as 'First Cut Rules'. By following these rules, the designer is able to produce a physical database structure which will at least work (though perhaps not very efficiently). Each separate DBMS or file handler has its own set of First Cut Rules, and Figure 12.3 gives an example of a set of such rules for an (unspecified) DBMS of the hierarchical type.

EXAMPLE SET OF FIRST CUT RULES

1. Ignore any 'unused' relationships; (ie. those which are not part of any transaction path).

2. Identify hierarchies by using one-to-many relationships, treating the 'one' entity type as the parent, and the 'many' as the child.

3. Any 'child' entity type which has more than one master must be allocated to one of those masters. As a rule, the master chosen should have the smallest average number of 'many' occurrences; (eg. a sales order will have a small number of order lines associated with it, whereas a stock entity will have thousands of order lines referring to it).

4. Any entity type to which there is an external access (ie. where any transaction path starts) must be at the top of a hierarchy.

5. Each of the hierarchies becomes a separate file.

6. Pointers and other file connection techniques must be used to support relationships which cross hierarchies.

Figure 12.3 First Cut Rules for an Unspecified Hierarchical DBMS

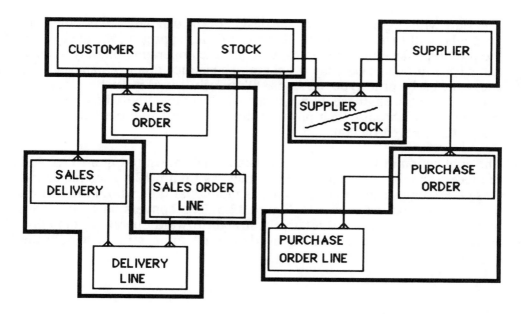

Figure 12.4 The Data Model split into Hierarchies
in accordance with First Cut Rules

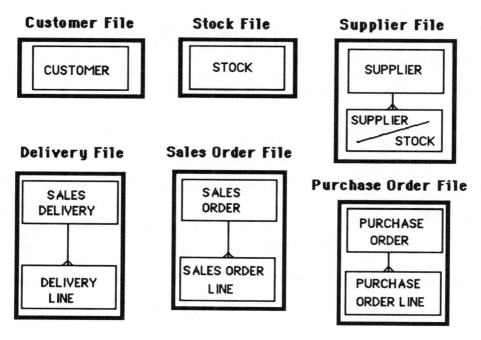

Figure 12.5 The First Cut File Design

The staff with responsibility for the design of the database are not the only ones who will make use of this First Cut Rules approach; the analysts in the prototype development teams, who are each operating on only a small part of the full database, may use them to construct 'dummy' databases for use with early prototype versions. Obviously there must be close co-operation between these two groups of designers, but they may not completely synchronise their physical structures until after the full database design has been carried out.

The application of these rules to the data model shown in Figure 12.2 will produce a number of separate hierarchies, as Figure 12.4 illustrates, and these hierarchies can be put forward as individual files within the overall database context, as shown in Figure 12.5. The conversion of this data model into these files is a relatively straightforward operation, though a certain amount of judgement is needed in deciding how to handle the Supplier/Stock entity type.

1.2.2 Optimisation

Once the first cut version has been created, the database specialist designers are able to get to work on 'tuning' the database for high performance. They will make extensive use of the data usage figures, and will carry out a number of iterations on the proposed database, suggesting different methods of access, and combining entity types to improve response times. In fact it is not uncommon for entity types which were separated out as a result of *first* normal form checks to be re-combined with their 'parent' entity types, simply for performance reasons!

FILE USAGE CHART

FILE NAME	RECORD NAMES	FIELD NAMES					
SALES ORDER	Order	Order No, Order Date, Customer No, Credit Status					
	Order Line	Item No, Quantity					
Access Source	Type	Process Id.	Frequency	Records Accessed	Access Time	TOTAL TIME	
Order No.	DA	Take Order	150 per day	15	45ns	6750ns	

Figure 12.6 A File Usage Chart

Clearly, in order to be able to perform this optimisation the designer needs to have expert knowledge of the appropriate DBMS. However, the basic process of trying out a possible physical option by deciding on a file structure, then testing the proposed 'traffic' on that structure, is common to all DBMS design.

In Figure 12.6, an example of a File Usage Chart is given. This is the physical equivalent of the logical Data Usage Chart. The designer is able to estimate and record the expected accesses on a particular physical file proposal, and take decisions based on the results. The example given here is, again, for a hypothetical DBMS; the designer needs to adjust this chart to suit the particular file handler being used for the system.

1.2.3 Security and Integrity Considerations

An important aspect of the database design is the incorporation of the necessary security and integrity checks within the database itself. The requirement for these may have been identified during the early Business Analysis, but will have been formalised during the 'Identify Necessary Controls' step of the detailed process design. Some of the checks can be carried out within the computer processes, but often the best approach is to build them into the database definition. To what extent this is possible depends on the DBMS / 4GE in question, and also on the skill of the database designer. It can be a highly complex and technical task, and the quality and experience of the design staff can make all the difference between success and failure. Whatever the effort required, it is imperative that inaccurate and inconsistent data is not allowed to corrupt the database, and that unauthorised access to database information is prevented.

2   COMPLETING THE PROGRAM DESIGN

Just as with the process of designing the physical database, the detail of this activity is heavily dependent on the software to be used. As a result, it is not possible to be specific about the best techniques to use.

The process itself is equally dependent on the type of business system being developed, and on the analysis & design approach taken in the earlier stages. It can however involve two different types of activity, each of which will be examined briefly. These are:

1. Adjusting the prototypes to cater for the optimised database

2. Completing program specifications for construction using a traditional 3GL approach.

2.1 Optimising the Prototype

This involves taking information from the database designers on the finally agreed structure and definition of the database, and adjusting the 4GL code in the various prototypes to enable them to handle the changes. Hopefully, in most cases this will simply be a matter of 'tidying up' the code, but there may be some circumstances where major re-organisation of code is necessary. Analysts may sometimes find it worthwhile to re-draw the path analysis diagrams, this time to illustrate the physical rather than logical paths.

In extreme cases, this optimisation process may involve the re-writing of a prototype in a third generation language for performance reasons: (situations where this type of approach is necessary should become rarer and rarer as 4GEs develop and improve).

When the prototype has been revised, it can be issued as a new version (referred to as the 'Optimised' prototype). In terms of the user interface, this version of the prototype should be EXACTLY THE SAME as the previous version; the optimisation changes should be transparent to the user. This means that the process of exercising the prototype will involve a different approach. The user in the development team is simply required to re-apply all the tests carried out on the earlier version(s), to check that no corruption has been introduced.

Once the Optimised prototype has been proven, that particular part of the system is considered to be complete, and the next stage is to implement it as a live system. This obviously can involve a whole series of activities, such as the link testing of different system components, the set-up of a master file of current data, the training of staff, the formal acceptance of the system by user management, the formal change-over from the old system, etc.

2.2 Traditional Construction

In spite of the strong emphasis on prototyping in this methodology, it is recognised that in many business computer systems there are massive complex calculation processes, for which the use of 4GLs and prototyping are currently unsuitable. In these circumstances, the programming would have to be carried out using rigorous structured programming techniques, and by a specialist programmer. Such programming techniques are obviously not included within the methodology, but the production of the *process description*, the specification from which the programmer is to work, must be part of any general analysis and design approach. Within Systemscraft, the process descriptions for non-prototype components are constructed using the same basic design steps as have been described for the development of prototypes.

A 'prototype' can be considered simply as a specification of physical requirements. It is written in a fourth generation language, which can be understood directly by the computer. It is however just as valid to write that specification in another language, for example 'Structured English', which can be understood by a 3GL programmer, who can then translate it into something that the machine can understand. This means that within the Systemscraft design stage, wherever the building of a prototype has been mentioned, a text process description can be constructed instead.

Unfortunately, whereas a prototype can be tested and validated by the user while it is being developed, for non-prototype specifications, a whole series of extra activities are required. These include the coding of the computer instructions, the detailed program, system and user-acceptance testing, etc.. Descriptions of how these activities are organised and carried out are included in most of the more traditional books on systems analysis and design.

3   CONCLUSION

This chapter contains a somewhat over-simplified account of the processes of designing a database schema, and adjusting the prototyped computer programs to incorporate its structure.The exact activities involved are so dependent on the hardware/software environment that only a brief and general appreciation of the task can be given here.

Having said that, it is necessary to stress the importance of good physical design. The whole purpose of taking a logical approach to data analysis is to identify the 'potential' inherent in the data, and to make it easier to evolve the system towards making use of that potential during its lifetime. Poor physical design can close down this potential.

13 ADDITIONAL ASPECTS OF THE METHODOLOGY

So that completes the detailed description of the steps, stages, tools and techniques which together make up the Systemscraft methodology. All that is necessary now is to examine briefly a number of general topics which do not fit within the previous chapter headings, but without which the coverage of the methodology would be incomplete. These topics are:

1. Documenting the Systems Development

2. Maintaining a system developed using the methodology

3. Some conclusions on Systemscraft

1 DOCUMENTING THE SYSTEMS DEVELOPMENT

Earlier in the book the format of the 'Business Requirements Specification', the formal documentation for the Business Analysis stage of the development process, was described. Here, at the end of the 'Systems Design' stage, it is necessary to provide a similar description of the structure and content of the 'Systems Design Specification'. There are however various other types of report, manual, file and specification, together making up the full set of documentation for the proposed new system, and before embarking on an examination of these, it is worthwhile discussing the different kinds of documentation that exist and what purpose they serve.

1.1 The Purpose of Documentation

The purpose of documentation is communication. The analysts involved in developing the system need to communicate with a number of people before, during and after the analysis and design process discussed here. The information obtained needs to be recorded in a format which facilitates easy access and retrieval. The results of the analysis activity, and the ideas considered during the design stage (both those accepted and those rejected) also need to be captured in some form of text, firstly to help in the completion of the development process, and secondly to support the running and the maintenance of the system when it goes live.

279

There are basically two forms of documentation. These relate to the two groups of people involved in the development, and their different information needs:

1. The **Users**. (The term is used here to include the managers, owners, and the people involved in running the system.) The documentation for these people must be formally prepared by the development team (some of whom are themselves users). This documentation is considered to be included among the 'deliverables' of the system. In the Systemscraft methodology, the only documentation deliverables thought to be necessary are the following;

   > Business Requirements Specification

   > Systems Design Specification

   > User Manual

   > Operating Instructions.

2. The **Developers**. (The term is used here to include the analysts, designers, prototypers, programmers, project manager, database specialists, etc. involved in the development process. It can also include some of the users who are heavily involved.) The documentation for these people consists of all the relevant information built up about the system throughout the duration of the study. This is commonly known as the *Working Papers File*.

1.2   The Working Papers File

The working papers file is the analysts' complete set of information relating to the project. It is the source from which the formal user documentation is produced (though these user documents themselves also become part of the file of working papers).

This file must be organised in such a way as to make it possible for every piece of information relating to the systems development to be stored and retrieved as effectively as possible. The structure of the file has to be loose enough to allow for any type of information relevant to the system (and this can vary greatly, depending on the nature of the system). At the same time the structure must be sufficiently well defined to provide clear indications as to where the different types of information should be stored, or from where they can be retrieved. This should be accomplished with the minimum of data duplication.

This is an obvious area in which some form of computerised text (and graphical) retrieval system would be of great advantage. One is tempted to think in terms of a CASE tool, but the majority of such tools are concerned more with the graphical development of structured models. They clearly have a major part to play in maintaining formal documentation, but they are unable to handle the enormous quantity and variety of multi-faceted information which accumulates into the Working Papers File. Modern research suggests that **Hypertext** systems are most likely to provide the solution. As Neilsen points out in his book, *Hypertext and Hypermedia*, there are several such systems already in existence. However, it seems that a well-proven commercially-available Hypertext system for maintaining the full working papers for a systems development project is still some years away!

At this time, when the whole process of systems development is undergoing major revision, it would be unwise of me to suggest a *best possible structure* for a working papers file. This is likely to differ from organisation to organisation, and from project to project. Systems Development Managers who are trying to identify standards and guidelines for this would do well to examine the recommendations put forward as part of the NCC Basic Systems Analysis course; (this is well documented in NCC publications).

Alternatively, small sections or teams of analysts who are looking for a way to support a 'Systemscraft' type of development approach may consider using the chapter headings from parts 2 and 3 of this book as basic classification categories in which to store their gradually accumulating material.

1.3   The Systems Design Specification

Just as at the end of the business analysis stage a formal report known as the 'Requirements Specification' must be produced (chapter 8.3) when the systems design has been completed, it is necessary for the team to issue a formal description of that design in a report known as the 'Systems Design Specification'. These two specification reports should have a common basic structure, in that one describes the 'logical' view of the new system, in terms of what the user requires, and the other describes the physical manifestation of that system.

Not only are the specification structures similar, but the actual modelling tools used for the equivalent components are closely related. This can be observed by comparing the diagram illustrating the two structures; (see Figure 13.1).

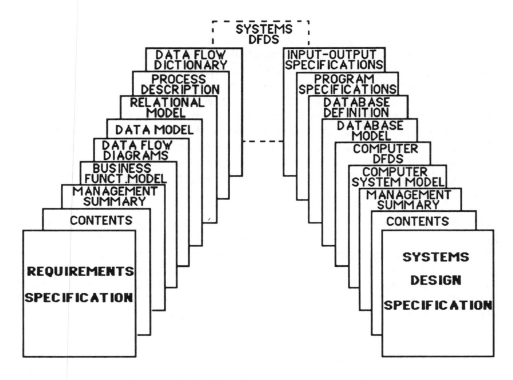

Figure 13.1 The Component Parts of the
Business Requirements and Systems Design Specifications

Apart from the contents sheet and the Management Summary, the Systems Design Specification is made up entirely from models built during the full design process.

The first of these is the **Computer System Diagram**, which is the hierarchical description of the suites, sub-systems, programs and modules in the new computer system. This is the physical equivalent of the Business Function Diagram from the Requirements Specification (though it has to be stressed that the computer system structure will not necessarily be based on logical function).

The second model, the **Computer DFD**, illustrates the physical data flows within the different levels of process in the system, and is required to conform with the Computer System Diagram in the same way that the Business DFD relates to the Business Function Diagram.

The next two components are concerned with the design of the physical database for the new system. Firstly there is a model of the physical database structure (the format of which will depend on whether a hierarchical, network or relational file handler is used). Secondly there is a definition of all the fields relating to each of the database files. Again these can be seen as the physical equivalents to the logical entity and relational models from the Requirements Specification.

The **Computer Process Descriptions** are the program and module specifications, documented to whatever level is considered to be necessary. Again this level will depend on the particular 4GL software being used. It has already been suggested that some of the programs may require to be written in a 3GL for performance reasons; in such a case, the program specifications may need to be in a very detailed form. Whatever form these specifications take, they should be supported in an appendix by the code listing.

The **Input/Output Specifications** refer to the computer screens and reports. (Note; they do not refer to clerical documents which are to be keyed into the system. These would be held separately in the 'User Manual'.) There is deliberately no method of modelling computer-user dialogues in the Systemscraft methodology; this is to try and enforce a simplicity of approach and a set of common-sense standards on the production of such dialogues. However, there will be some projects where the modelling of complex dialogues is essential. Where this happens it is recommended that a charting method based on the Jackson Structure Diagram is used. (This has the advantage of being very similar to the technique used for Entity Life History modelling.)

It should be noted that a minimal set of models is included in the specification. Diagrams like the Entity Life History and the Logical Path Analysis Diagram may be vitally important in the development of the design, but they are not needed to show how the finally agreed system is constructed. Such models would remain as part of the working papers file.

There is one important exception to this rule. Although the **Systems DFDs** are not strictly part of either the Requirements Specification or the Systems Design Specification, they are an essential link between the two. When the new system has to be maintained, any proposed amendments must be examined at the logical as well as physical level, and the only connection is via the Systems DFD. It is therefore suggested that the Systems DFD model be held in the Business Requirements Specification, replacing the previous 'Physical Requirements' component; (see Figure 13.1).

2   THE MAINTENANCE OF DEVELOPED SYSTEMS

The traditional view of systems development is that programs are built to handle the business requirements as they are seen at a particular moment in time. As time passes, changes must be made to these programs, to fix errors in the original design, and to include new requirements. These changes are 'patched' into the programs (albeit in the most appropriate places that can be found!). Eventually, when so many patches have been applied, the basic structure is so undermined that it becomes almost impossible to incorporate any more; the computer system is no longer a recognisable representation of the business requirements, and it is necessary to investigate, analyse and re-design the system from scratch. This is the logic inherent in the systems development life-cycle.

The more modern view is that systems and programs evolve throughout their lives. The changes which are made over time should not cause the structure of the computer system to 'deteriorate', but to improve! The system is after all moving towards the achievement of its potential. The changes that are made do not simply involve an alteration to the code, but automatically include a re-assessment and re-specification of the requirements.

The source of this change in philosophy lies in the modern concepts of structured systems modelling, database architecture and fourth generation development. It relates directly to the difference between **what is required from a computer system**, and **how that requirement is provided**.

The traditional view was born in the era where batch processing was dominant, and data was held on magnetic tape in sequentially organised files. Most of the complexity of such systems stemmed from the problems of creating a solution to standard and relatively simple business problems using a primitive, unwieldy and unfriendly hardware environment! Complex batch procedures had to be introduced, sort programs written, and temporary files constructed. Because of the nature of batch processing, much of the activity which is performed better by human beings than by computers was forced into the all-embracing computer system, making it even more complex and bug-ridden.

As a result of modern developments in hardware, software and systems design techniques, we are moving closer and closer to the situation where the structure of what is required from a computer system is the same as the structure of what the system provides; (to use a well-coined phrase, 'what you see is what you get'). The on-line interactive facilities, the automatic conversion of 'specifications' into computer-understood code, the use of object-oriented techniques to map the 'real world' into the system, all play their part in narrowing the gap between **requirement** and **provision**.

This means that amendments to the system during its working life can often now be expressed in the form of logical requirements rather than as intricate physical code. As a result, these amendments should cause no deterioration to the system structure, and therefore it should no longer be necessary to redesign the whole system completely every six or seven years!

What this means in terms of the Systemscraft methodology is that the process of changing the system after it has 'gone live' is basically the same as that for its original development. The working system can be considered as the prototype, and the users will discuss the required amendments with the analyst in the same terms as they would were they exercising the prototype. Amendments can then be applied, and new iterations can be issued or versions released. This is obviously an over-simplification of the whole process of amending live systems; there needs to be some form of change-control, possibly even of project management. All that is being stated here is the principle that Evolutionary Development continues throughout the life of a system, and that **there is relatively little difference between original development and maintenance when using an EDM.**

Having said all this, there will always be some situations where it is necessary to analyse, design and re-develop from scratch a system which no longer matches the user's requirements, even though the original development was carried out using the modern tools and techniques just discussed. The reason for this can be found in the difference between *continuous* and *dis-continuous* change. Charles Handy, in his book *The Age of Unreason*, describes this difference very clearly. For long periods in the history of any system, progress can be taken to mean 'more of the same, only better'. Sometimes, however, there is a sudden, often unpredictable, shift of direction, which negates or overturns many of the major concepts and components of the system. When the logic behind all or part of a system is undermined in this way, the only possible approach is to go back to first principles.

Bearing this in mind, the purpose of this examination of the evolutionary nature of change in computer systems has not been to deny the validity of complete system re-development, but simply to question its place as a regular and inevitable stage in the systems life-cycle.

3 A REVIEW OF THE SYSTEMSCRAFT METHODOLOGY

The Systemscraft methodology has been designed to provide a full set of modelling tools for the analysis and design of business computer systems. In the design stages in particular, there are a variety of tools, not all of which will be utilised in every system. Even when a modelling technique is to be used, there may be different levels of complexity and detail to which it can be applied.

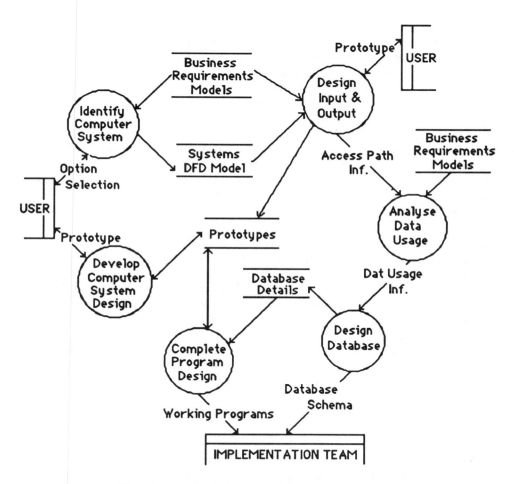

Figure 13.2 A Physical DFD of the Systems Design Stage

286

3.1 A Model of the Systems Design Stage

In an earlier chapter it was pointed out that, as a methodology contained tools for modelling systems, and as a methodology was itself a kind of system, then the tools could be used to model the methodology! At the end of the description of the Business Analysis stage of Systemscraft, a DFD model was used to illustrate the functions of the stage (Figure 8.10). In order to complete this model for the whole methodology, Figure 13.2 is used to illustrate the functions of the Systems Design stage.

This facility that the methodology has to model itself can be very useful to the senior analysts who are making decisions on how to tailor the methodology for a particular project. A DFD not unlike the one shown in Figure 13.2 could be used as the vehicle for communicating their decisions to the project teams.

3.2 Using a Minimal Set of Techniques

One of the guiding principles used when selecting modelling techniques for a methodology is that the minimum number of different types of technique should be included, and the same basic technique should be used at several different stages of the development for different purposes. This makes the methodology easier to learn, and facilitates the build-up of expertise with each technique.

In Systemscraft this approach has been adopted extensively. For example, the Logical Path Analysis technique can be adapted to illustrate the physical accesses to the new database system, the Data Usage Chart can be converted for use as a physical File Usage Chart, and a variant of the entity modelling technique itself is very often used to demonstrate the final physical structure of the database.

The most important example of a multi-purpose modelling technique within Systemscraft is of course the DFD, which is used

> to help analyse the business flows,
> to identify the computer system boundary,
> to help design the overall computer system structure.

The relationships between the functions and processes shown on the three versions of the model are critical, in that the designer who is maintaining a computer system must be able to identify the function which any one of the computer processes is created to serve. Figure 13.3 illustrates these inter-connections.

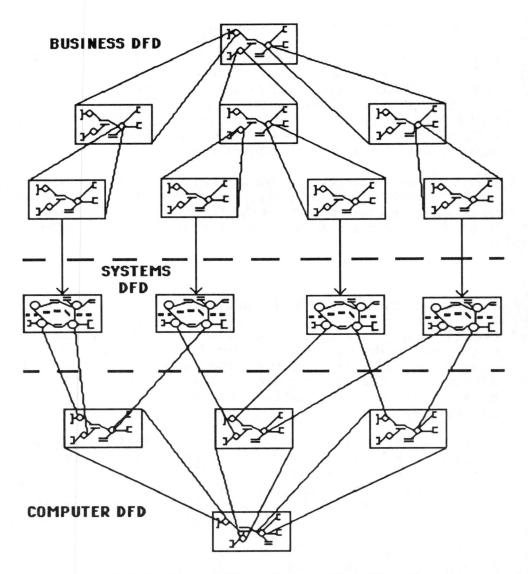

Figure 13.3 The Three Different Forms of Data Flow Diagram

3.3 Simplicity and Complexity

The full methodology as described in parts 2 and 3 of the book may seem very rigorous and off-putting to anyone who is only concerned with relatively small 4GE developments. The intention has been to illustrate that system size is not a limitation when deciding whether or not to use an EDM. It is of course recognised that still the majority of systems being developed using fourth generation tools are small or medium-sized, and that several of the techniques discussed in the book provide analysis and design facilities which are not needed for such systems. The flexibility and scalability of EDMs is discussed in some detail in part 4 of the book, but it is worthwhile here to emphasise that for most small-to-medium developments, **the basic minimum version of the methodology** will suffice. Figure 13.4 (which is a copy of Figure 3.5) illustrates this approach.

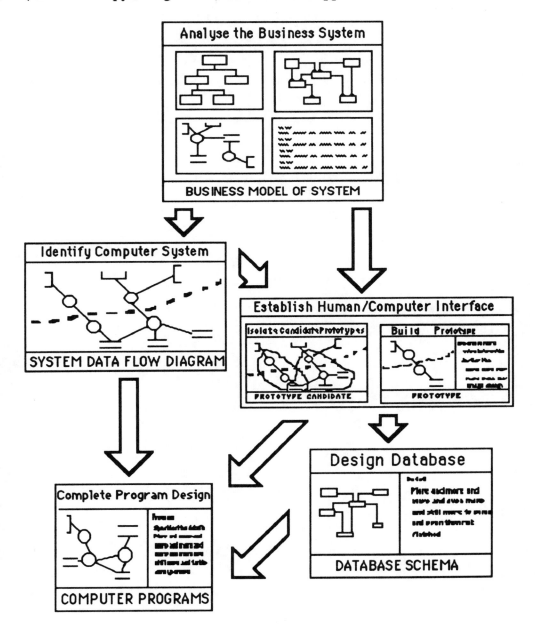

Figure 13.4 Minimum Form of the Systemscraft Methodology

Whereas under-utilisation of systems development tools can result in poor quality systems, their over-utilisation leading to waste of resources can be almost as bad. The analyst in charge of the development must use judgement in deciding exactly what techniques are to be used, and to what level they are to be applied, in order to balance the two risks. The propensity to facilitate this is claimed to be one of the main benefits of the methodology.

Part 4

14 TAILORING AN EVOLUTIONARY DEVELOPMENT METHODOLOGY

Part 4 of this book is concerned with pulling together the main ideas put forward in the rest of the book, and placing them within a management context. The achievement of this however necessitates a certain amount of **repetition of ideas** from earlier in the book. I make no apology for this; after all, the ideas in question are likely to be new to the majority of readers, and their implications need to be explained and justified.

This chapter is primarily concerned with the flexibility and scalability of an evolutionary development approach, and the ways in which an EDM can be 'tailored' to fit specific circumstances, the chapter after this one concentrates on the methods of project management for an EDM development, and the final chapter of the book is reserved for an overall summary, highlighting the essential components of any truly evolutionary approach.

An evolutionary development methodology must be able to handle the development of a variety of different types of systems, some heavily data-oriented, some with very complex processing, others with rigorous control requirements, etc. It must also be able to tackle systems of different sizes, ranging from the relatively trivial two-person-day project, through to the several-person-year major development.

There are in fact three different levels at which any general purpose evolutionary methodology may need to be tailored:

1. The Organisation level, to suit the culture and applications portfolio of the company

2. The Environment level, to take advantage of each different fourth generation environment used within the company

3. The Project level, to select the appropriate components of the methodology to be used for a particular project, and the correct degree of complexity with which to employ them.

It is the last of these three levels with which we are most concerned, but each level will be discussed in turn. Where appropriate, the Systemscraft methodology will be used to illustrate examples, but the points being made throughout this final part of the book relate to all evolutionary development methodologies.

1 TAILORING AN EDM TO AN ORGANISATION

At this moment, within organisations of varying degrees of size, systems development managers are considering whether or not to adopt an evolutionary prototyping approach to some types of systems project. It has been pointed out in earlier chapters that prototyping and the use of fourth generation languages have already proved themselves as effective in the development of inexpensive, user-friendly systems, but the problem is often how to make the approach fit with the company's existing standards and methods.

1.1 The Place of EDM within a Systems Development Strategy

An increasing number of organisations are adopting an approach whereby they provide three entirely different types of systems development environment for their customers:

1. A structured systems methodology of the 'cookbook' variety for building large and complex systems

2. A much simpler 'toolbox' type methodology for small and medium-sized systems (to be developed using 4GLs)

3. An Information Centre facility for end-user support.

Many of these organisations have also recognised that in order to take full advantage of the potential of the fourth generation of software development tools, the approach to be adopted for small and medium-sized systems should be some form of Evolutionary Development Methodology.

Furthermore, this smaller methodology, if it is to be an EDM, cannot be the result of a series of simplifying adjustments to the larger one; the inherent concept of the traditional systems development lifecycle must be jettisoned!

Obviously, the organisation must also have some mechanism for deciding which of these three environments to employ for each potential project. However, the most interesting point about this is that the criteria to be used in this selection process are changing very quickly. Originally there was a relatively small range of types of system considered suitable for a prototyping approach. As a result of improvements in 4GEs and CASE tools, and also through the build-up of experience in the use of such methods, the scope for the use of prototyping techniques has been identified in many more types of project. So much so that it has become easier to think in terms of the small range of systems where some form of prototyping is NOT appropriate!

The implications of this changing pattern are that, as time progresses,

1. Fewer and fewer developments will be carried out using the large 'cookbook' third generation structured methodologies

2. The 4GE-based prototype-oriented evolutionary development methodologies will be used for more and more of the larger types of system

3. Equally, the number of Information Centre developments will increase, to include many of the types of system which are currently handled by the simpler prototyping methodologies.

Eventually, it is expected that methodologies which incorporate some level of evolutionary prototyping as a major component will become the most common form of systems development. There are two major reasons for assuming this. The first is that as more and more fifth generation ideas are made use of in the business arena (in the form of knowledge-based and expert systems) the evolutionary approach will have to be given more credence: there is really no other way of developing an expert system!

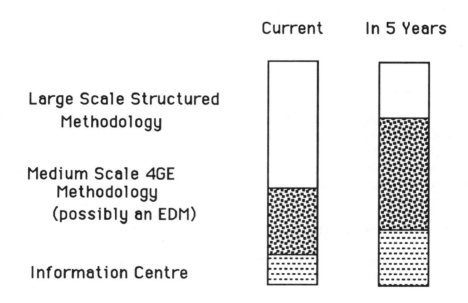

Figure 14.1 A Prediction for Future Systems Development

The second reason relates to the fact that, whereas there are major difficulties in the inclusion of evolutionary constructs within a 'pre-specification' type methodology, there are really no problems in incorporating the traditional life-cycle development within an evolutionary development methodology. In such methodologies, the 'prototypes' which are evolved are simply coded versions of process specifications; one could follow exactly the same methodology steps, and produce formal program specifications instead of working models. This means that, in theory, a methodology like Systemscraft could be used to replace existing life-cycle methodologies for traditional forms of development as well as for evolutionary ones.

1.2 Adoption of an EDM within an Organisation

The process of adopting some form of evolutionary development approach within an organisation, involves a great deal of careful planning, and consist of four basic steps (Figure 14.2):

1. A Survey of the Market Place

2. Procurement and/or Initial Development

3. Introduction through Pilot Projects

4. Provision of Support and Expertise

1.2.1 Market Survey

Senior staff within the IT division need to talk to owners of methodologies, to find out what is available. They must also discuss with other organisations who make use of such methodologies, what is being used, and how effective it is. From evidence gathered in this way, they should be able to firm up on a list of requirements on which they can base their selection.

This list of requirements is likely to include the following:

1. The essential constructs of a practical EDM; (a list and full description of these is given in Chapter 16)

2. Some level of compatibility with the systems development methods currently in use within the organisation; (to enable it to fit into the SD strategy, and to simplify the migration)

3. The availability of sufficient and adequate training, and of consultancy support during the early stages of the methodology's introduction

4. Compatibility of the method with the one-or-more 4GEs and CASE tools to be used within the organisation; (it must be flexible enough to be tailored to fit with them all)

5. Suitability of the method for the nature of work carried out by the organisation; (eg. if there are time-critical or security requirements in many of the company's systems, the method must include components which can deal with them).

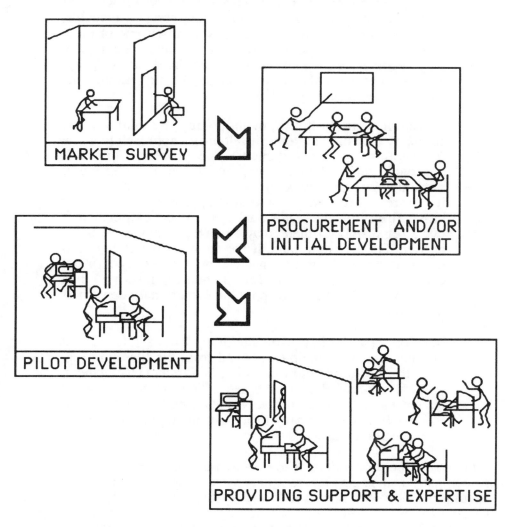

Figure 14 .2 Adoption of an EDM within an Organisation

1.2.2 Procurement and/or Initial Development

The list of requirements identified by the analysts who are investigating the feasibility of an EDM approach will be used initially to help the organisation decide on its general approach.

There are three main options which they must consider:

1. Obtaining an 'off-the-shelf' methodology (eg. Hoskyns' '4GDM', Southcourt's ESM or a variant of Systemscraft) and either using it in its standard form, or tailoring it with the assistance of the suppliers. There can be major advantages in taking this approach. For example, there will always be external training facilities and consultancy services available. However, there may also be dangers, in that some of these methodologies provide limited flexibility, and the company may become locked into one particular hardware/software environment. Also, the initial costs may be high (though the benefits should quickly follow).

2. Developing an 'in-house' methodology, by adapting the company's existing systems development approach to allow for evolutionary development. The problems that this approach faces depend very much on the particular methodology currently being used within the organisation. If it is a relatively strict 'cookbook' approach (eg. JSD) or based firmly on the traditional systems development life cycle (eg.SSADM) then major concepts may have to be revised. On the other hand, if the methodology being used provides some flexibility in the type and sequence of its component techniques (eg. 'Information Engineering') then it will be much simpler to apply the evolutionary constructs discussed in this book.

3. Developing an 'in-house' methodology from scratch, based on evolutionary principles derived from the literature (eg. from this book, or from Gane's *Rapid System Development*, or Vonk's *Rapid Prototyping*, etc.). This can be a hard but rewarding path, in that it involves learning by mistakes, but it does allow the company to 'evolve' an approach which is very specific to its needs.

The choice of option will depend on a number of characteristics related to the company's ethos, its view of future systems development needs, and its existing position with regard to formal systems methodologies. However, we can assume that some level of tailoring will be necessary, even if (as in the case of the first option) it may be carried out by consultants.

Here are some examples of the kind of adjustments that might need to be made to an evolutionary development methodology to tailor it to the specific needs of a particular organisation. In several of these examples, Systemscraft is assumed as the 'prototype' methodology, though the same principles would apply to any other framework chosen.

Firstly, the standards and symbols used in some of the EDM's modelling techniques can be altered to fit in with the techniques already being used in existing in-house methodologies.

For example, if a Yourdon methodology is already in use for the larger systems developments, then it may be wise to include the 'State-Transition' diagramming technique in place of the 'Entity Life History' approach suggested here for event and behaviour modelling. The two techniques have similar purposes, and there are advantages in standardising on the use of techniques across the different systems development environments.

Similarly, the Controls Analysis techniques put forward in chapter 11 may well clash with standards already being used as part of the organisation's internal audit procedures. Some compromise may have to be reached.

Both the cases above illustrate the idea that the 'physical' form of the proposed methodology may need to be reconsidered (as befits a prototype) but the 'logical' functions, eg. of examining the behaviour and controls, must somehow be taken into account.

Even when the basic technique for a particular function is not to be replaced, there may be some advantage in changing the symbols and layout in order to fit with existing company standards and experience. For example, if the existing methodology for the development of large systems is a variant of SSADM, then there is a strong case for changing the shape of the various DFD symbols used here to coincide with those used for the larger methodology. However, care needs to be taken when making such a change, because the philosophy underlying different methodologies' interpretation of what is apparently the same technique can be confusing. It is for example generally recognised that SSADM takes a less business-oriented view than many methodologies, preferring to concentrate on the information system to be computerised. As a result, the 'logical' DFDs produced in SSADM are often considered to be more 'physical' than those of some other methodologies; (eg. they only include parts of the system which are to be computerised). So, a straight replacement of DFD symbols and standards by those of SSADM could result in the loss of some important evolutionary traits.

A similar but less critical situation might arise if the organisation made use of one of the important range of methodologies which incorporate the

'Chen' approach to logical data modelling, rather than the much simpler form which has been put forward in this book. The best known of these is the Yourdon methodology, but there is also the French national methodology, 'Merise', which has a strong following in various parts of Europe. There are some apparently striking differences in the notation used in these two alternative ways of arriving at the data requirements, but in fact the solution from either approach is likely to be more or less the same.

Perhaps the most important criteria affecting the tailoring of the EDM at this level are those relating to the kind of systems which the organisation is required to develop; ie. its 'application portfolio'.

For example, if the EDM is only to be used for the development of small systems, then perhaps the absolute minimum form of the methodology (eg. as in Figure 3.15) should be put forward as the standard version, and the more complex aspects can be treated as 'add-ons', only used for certain types of system. Obviously the guidelines for identifying these complex systems would need to be drawn up carefully.

Another situation which might necessitate a major alteration to the proposed EDM is one where the organisation has adopted a Corporate Data Strategy. Normally this means that a special team has been set up to create, control and oversee one large logical corporate database for the whole organisation. This team would therefore have responsibility for the building of all entity and relational models, a task which is usually included within the evolutionary development process. In these circumstances, there is clearly an extra layer of communication required, and it may be necessary to re-order the steps in the evolutionary development method to allow the completion of the data modelling process before prototyping can begin.

It can be seen from the example cases mentioned above, that if a successful in-house methodology is to be devised, then a very flexible EDM framework must be used as a starting point.

1.2.3 Introduction through Pilot Projects

The simplest and best approach to introducing a methodology within an organisation is to use a series of relatively small pilot systems. Initially a small team of staff are trained in the use of the methodology, and they are assigned to tackle a number of non-critical (but equally non-trivial) systems projects. The intention is that they will :

develop skills in the use of the methodology

prove the basic principles of the approach (to themseves and to management)

identify those areas where the methodology needs to be adapted
to suit the organisation.

These early projects will be chosen because they fit clearly within the
selection criteria set for the use of the new methodology: (later projects can
test the scope boundaries). This last point is important, because every major
new change in practice must be 'nursed' through its early stages: evidence
suggests that the reason why so many new initiatives fail is that they are
brought into conflict with an unfavourable environment before they are
robust enough to stand the pressure.

1.2.4 Provision of Support and Expertise

Once these early projects have been completed, the members of the team can
be appointed to lead new teams, and in this way the ranks of experienced
practitioners will gradually swell. Some of these pioneers may eventually be
re-united to form a 'centre of expertise', a team who will :

provide advice and support in the use of the approach to staff
working on relevant projects

maintain and enforce the use of the methodology standards
within the organisation

revise and update the methodology and its documentation in line
with new developments within IT.

This idea of a 'research and development ' function within the support team
is particularly important at this time in the history of systems analysis and
design, when major concepts like CASE, object-orientation, and graphical
user interface are changing the nature of systems development almost day
by day!

2 TAILORING AN EDM TO A PARTICULAR 4GE

Many organisations make use of more than one fourth generation
environment within their systems development strategy. For example, it
would not be unusual to find 'FOCUS' being used alongside 'Sourcewriter'
and 'Dbase 4', perhaps with a special statistically-oriented 4GE like 'SIR'.
The reason for this is that each 4GE was originally assembled with specific
objectives in mind, usually to simplify the development of a particular type
of system. As time has passed, all major 4GEs have been revised and

extended to encompass most of the common types of business system, but they each have their strengths and weaknesses, and the discerning analyst/designer should be able to adopt the most appropriate with which to solve the problem.

Where such a situation exists, there are definite advantages in producing slightly different versions of the company's EDM for each of the environments, to take full advantage of their individual strengths.

Usually there will be little need for adjustment to the Business Analysis component of the methodology; this tends to address the logical requirements, and specifically excludes the physical environmental aspects. However, there are a number of circumstances where some adjustments to the methodology standards may be beneficial. One obvious instance is where the 4GE incorporates a CASE tool component (as in the case of FOCUS, Oracle, ICL Quickbuild, etc.). The changes may involve something as simple as replacing the 'delta' symbol (which indicates a 'many' relationship on the entity model) by a double-arrowhead, or something much more complex, such as substituting a different form of process modelling for the DFD. Clearly, the advantages and disadvantages of such changes must be weighed up carefully.

It is of course the Systems Design stage of the methodology where changes to the methodology are most likely to be beneficial. The two most obvious areas are those that address the set-up of the physical database, and the conversion of the logical process specifications into fourth generation code. In both cases, the movement from the logical to the physical can be simplified if the particular modelling techniques used lend themselves to easy translation to the specific 4GE constructs.

One example of such a change is the assignment of names to the relationships on an entity model, when the proposed development environment makes use of a Codasyl or network database management system. In such an environment, each entity relationship may have to be set up as a named index, and the use of named relationships on the original logical entity model simplifies this task, as well as establishing a clear continuity between the logical and physical versions of the system.

Similarly, as different 4GEs allow the use of very different styles of language statement for building programs, there may be advantages in adjusting the logical process description techniques, or the transaction path analysis models, to fit more closely with the particular language to be used.

The process of making decisions on the tailoring of a methodology to take advantage of the strengths of different 4GEs is a difficult one, because a critical balance must be maintained. The most important requirement is to have a company-standard for the methodology, one which can be specified, monitored and controlled, and any variation on this standard to suit an individual 4GE complicates the overall set of standards. Such variations must be carefully justified and formalised.

3 TAILORING AN EDM TO A PARTICULAR PROJECT

The key to the success of a modern methodology is its flexibility and scalability; ie. whether or not it is able to cope with the ever-increasing variety of types of system, and the extensive range of possible system size. Some methodologies approach this by providing a fixed framework of phases and stages, and a cookbook of instructions on how to·complete each stage. The more modern approach is to conduct a small initial investigation, and then 'custom-design' the development methodology for the particular one-off circumstances of the project. This approach is essential with evolutionary development.

This initial investigation normally constitutes a major part of the Feasibility Study. Feasibility studies will be discussed in more detail in the next chapter, which deals with the subject of project management. As one might expect, not only are we concerned with what activities are to be carried out,· be we need to plan and estimate how long they will take, and how much resource they will require. However, at this point we are only interested in how decisions are made concerning the method of development.

There are great advantages in using a strong methodology **framework** as a kind of template to guide the 'tailor' through the steps in the development process, checking whether each is necessary for this project, and if so, how it should be approached. During the following discussion the Systemscraft framework will be used as an example. The framework is shown in full detail in chapter 3 (Figures 3.2 and 3.3), but the different components will be illustrated here in this chapter as they are examined.

3.1 Whether or not the EDM is to be used...

The first decision to be made when a particular IT project is mooted, is which of the different systems development environments should be used for the job. Earlier it was suggested that there might be three alternative environments, one of which catered for an evolutionary development approach. Some of the main criteria for the selection of the most appropriate environment are:

1. the size and nature of the task

2. the staff and other resources available

3. the importance of the task, and its urgency

4. the attitude of the users (managers)

5. (not least) the capabilities of the EDM being considered

This decision is unique for each project being examined, and depends entirely on the situation and circumstances of the organisation at the time. Figure 14.3 illustrates how the scope of systems which can be dealt with by each of the environments overlaps considerably: very commonly a particular project could be carried out using either a traditional structured methodology or an EDM. This being the case, the decision as to which approach is to be taken may depend heavily on the availability of staff and resources, though it should be borne in mind that the evolutionary option is likely to provide the cheaper, quicker and better solution.

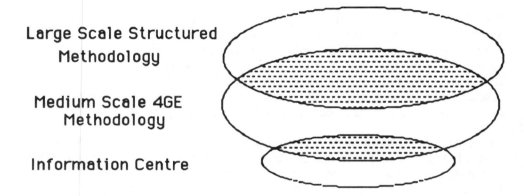

Large Scale Structured Methodology

Medium Scale 4GE Methodology

Information Centre

Figure 14 .3 The Scope of the different Development Environments

3.2   The Initial Investigation

Before any further decisions can be made, basic information about the system must be collected and analysed. This must be carried out quickly, and by skilful and experienced staff. For example, as part of a Systemscraft feasibility study, an average of three-to-four person-days would be set aside for the task, and it would be carried out by two senior analysts. The information would be collected almost exclusively from managers rather than operatives; it is the high level requirements that are sought here rather than the detail of the existing system.

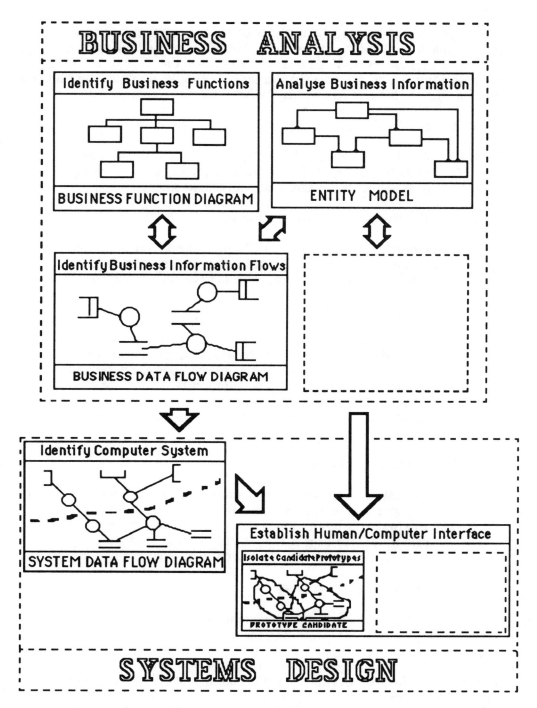

Figure 14.4   Models built during the Initial Investigation

The approach adopted for the analysis of the information is the same as in the two main stages of the methodology, only it is carried out more quickly, and at a higher level. Figure 14.4 illustrates the components of the methodology which are used in the task. It should be noted that all the models relate to the *required* system rather than the *existing* system.

1. Identify Business Functions

   A full Business Function Diagram is constructed, down to its lowest layer. This is done as quickly and accurately as possible, and is checked with management.

2. Identify Business Information Flows

   Only the lowest level DFDs are completed here (for maximum speed). A context diagram may be used to help check that all external agents have been considered.

3. Analyse Business Information

   A preliminary Entity Model is built using intuitive methods, as discussed in chapter 6. Again this is done quickly, with emphasis on the most important *asset* and *transaction* entity types. The numbers of entities for each table are estimated, and a rough list of attributes assembled (but no formal normalisation would be carried out).

4. Identify Computer System

   The initial proposal for the Computer System Boundary is put forward on Systems DFDs (derived from the Business DFDs), and is discussed and agreed in principle with user management. This is a particularly critical stage, involving the selection of technical options.

5. Isolate Candidate Prototypes

   Candidate prototypes are roughed out and marked on the Systems DFD, providing the estimators with an idea of the project scope.

The analysts who are carrying out the Feasibility Study will also assemble a list of the user's physical requirements and constraints. This, and the information gleaned from the above models, should be sufficient to enable them to consider how the methodology should be tailored. On hopefully rare occasions, some information picked up during this study will highlight the need for a much more extensive examination before feasibility (and tailoring) decisions can be made.

The importance of this initial investigation, and decisions taken as a result of its findings, cannot be over-estimated. The best possible analysts and the most experienced, knowledgable and committed users should be assigned to the task. It may only last for two days, but they are likely to be the two most critical days of the project!

3.3 Tailoring the Development Process

There are a whole range of decisions to be made as to which modelling techniques are to be used during the development stages of the system. Again, the framework of the methodology can be 'walked through', and each stage examined in turn.

3.3.1 Pre-Business Analysis

The first decision to be made is whether or not it is necessary to spend time building **Physical DFD models** of the existing system. There are two circumstances where this is beneficial:

1. where knowledge about parts of the existing system is fuzzy, and there is seen to be a need to model the existing business area activities in order to understand the implicit requirements. This should rarely occur , and when it does, the physical model produced is simply used to assist in the logical analysis. It is not included in the formal documentation of the system.

2. where there are short term gains to be realised by 'tweaking' the existing system, thereby improving its performance during the period while the new system is still under development.

In either case, estimates should be made for the time required for the building of these models, and the carrying out of the associated tasks.

3.3.2 Business Analysis

In all normal circumstances it will be necessary to produce all four modelling components of the Business Analysis process (ie. BFD, DFD,

Entity Model and Relational Model). All systems, irrespective of their nature, involve both data and processes, so both of these aspects must be fully explored.

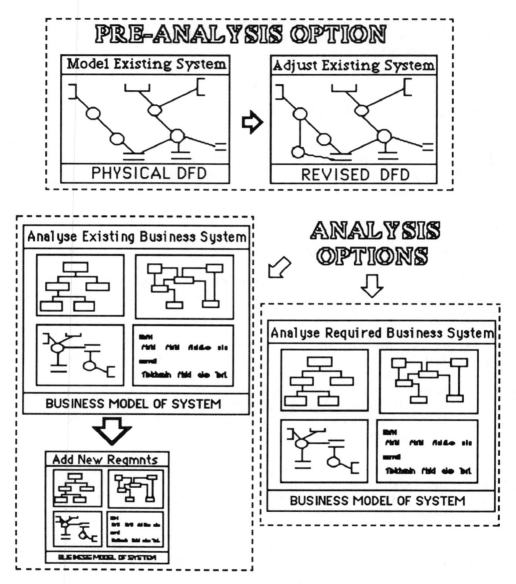

Figure 14.5 Business Analysis Tailoring Options

There is however a decision to be made as to whether the *existing* system should be modelled first, then new requirements added in, or the *new* system should be modelled from scratch as an integral unit.

This depends very much on how different the new and old systems are to be. Obviously if the whole business approach is to be revised, then there is less benefit in building extensive models of the old system, and there is a strong case for going ahead immediately with the Required Business Model.

On the other hand, in most projects, the vast bulk of the business requirements are inherent in the existing system, to such an extent that the user is unlikely to specify them directly. The terms of reference usually imply that all the functionality of the existing system plus certain additional facilities are required. In such circumstances, it is often essential to explore the existing system in some detail to establish these requirements, and it is usually easier to model these existing requirements before considering how the extra new requirements can be blended in.

In other words, unless the new proposals are very different from the old, AND the system to be developed is relatively small and simple, then it is advisable to build an Existing Business Model, then evolve it into a Required Business Model.

There are obviously points to be made about the different sequences in which the four models are to be built, and the order in which different functions from the BFD are to be progressed. These are considered to be part of project planning, and are dealt with in the next chapter; here we are simply concerned with the tailoring of the methodology framework, and the selection of the appropriate components for the task.

3.3.3 Initial Systems Design

Initial systems design encompasses the two development processes of 'Identifying the Computer System' and 'Establishing the Human/Computer Interface' (as shown in the systems design road map in Figure 3.3). Both processes are essential to the development of any system, so both must be included in all tailored versions of the methodology. However, there are a number of possible variations which may need to be applied to the standard EDM to cater for different situations.

In fact, although it is assumed that the overall project being developed is suitable for an EDM approach (otherwise a different approach would have been taken), it is quite likely that some parts of the project may require some other form specification to that of an evolutionary prototype. At this stage, decisions have to be made as to which parts of the project will require which of these types of approach.

The options are:

1. traditional batch-oriented development, possibly with 3GL code (perhaps for performance reasons)

2. the use of package solutions (either completely 'off-the-shelf' or amended to requirements)

3. end-user development, with the users building parts of their own systems (eg. enquiries using some kind of high level language).

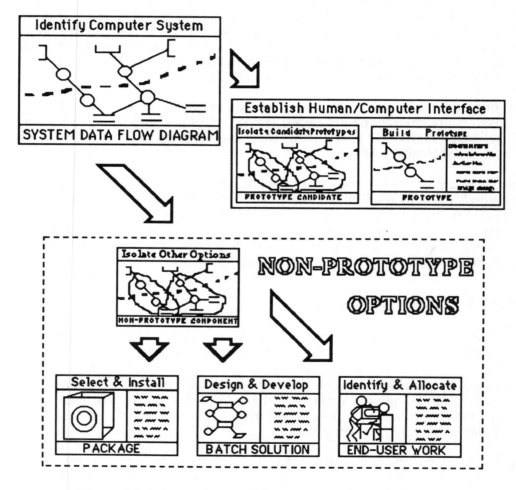

Figure 14.6   Early Systems Design: Tailoring Options

More will be said later about how project components which are developed in any of these ways are planned, and how they are incorporated in the overall development. It must be clear however that some **extensions** must be made to the standard EDM toolbox to handle these different approaches.

For example, where there are batch sub-systems to be developed, it may be necessary to make use of Sequence Mapping and Computer Run Charting techniques (used extensively in standard life-cycle methodologies). Similarly, some of the Information Centre procedures and practices may have to be introduced to assist with the end-user development.

In the case where a package solution is being considered, the existing EDM techniques of the Business DFD and the Systems DFD can prove to be extremely useful. Each of the potential packages can itself be modelled to show its functionality (Business DFD) and its user-computer interface (Systems DFD). These models of the package's facilities can then be compared with the same models showing the requirements of the system, and in this way, judgements can be made as to the most appropriate package to choose, and the amount of tailoring needed to be done to the package to satisfy the full set of requirements. When tailoring is needed, this may have to be done by the package owners. However, in some circumstances it may be possible to treat the current version of the package as a prototype, and develop it in an evolutionary way, along with the other parts of the project.

3.3.4 Detailed Design of the System's Database

Up to this point, most of the tailoring options have involved extending the methodology to be able to handle extra (and at the moment relatively exceptional) requirements. In the later stages of a methodology, it is more likely that modelling techniques exist for the more complex aspects of design, and the tailoring operation will involve simplifying the development process, by leaving out techniques which the particular project does not require. It should also be noted that the adjustments made to the methodology to take advantage of facilities provided by a specific 4GE are likely to apply more directly to these later design stages.

There are two stages involved in the detailed design of the database, as shown in Figure 14.7. The first of these, **Analysing Data Usage**, itself involves three different tasks, each of which makes use of a particular modelling technique. The tasks are:

1. To identify for each process or transaction, the data stores which must be accessed, and the sequence in which these accesses must take place. The technique used is the *Path Analysis Diagram.*

311

2. To map all of these transaction access paths on the master data model, thereby identifying the pattern of data 'traffic' within the system. The modelling technique used is known as a *Navigation Model*.

3. To examine the volume of access to each of the individual data stores, and provide detailed information to assist in the design of the physical file. The model used here is known as a *Data Usage Chart*.

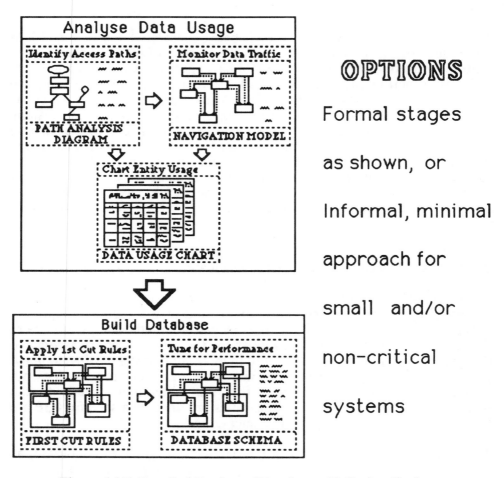

Figure 14.7 Detailed Design of Database: Tailoring Options

This formal process of analysing data usage only needs to be carried out where the system being developed is large, and has potential problems of speed of access. Alternatively, the problems may relate to complexity of access, where it is necessary to prove that certain information-accessing activities can actually be carried out! Obviously there is always a need for an informal analysis, if only to ascertain that no problem exists. However, it is essential for the analyst to bear in mind the possible expansion in use of the system once it goes live; often the initial system performs well, but once the enthusiastic users start to take full advantage of it, the response-time takes a nose-dive!

There may always be a case for carrying out the first of the three tasks, as this should be part of the process of designing each individual prototype. The modelling is carried out by the prototyper, and the path analysis diagram itself may serve as a kind of high level flowchart to help in the planning of the 4GL code.

The path analysis diagrams are then passed to the analysts who are dealing with the design of the database, and they are the ones who carry out the detailed usage analysis. Again, if a corporate database is being made use of, then some of this work may be done outside the boundary of the specific project, and therefore may not need to be considered by the tailor of the methodology.

One would anticipate that for most small and medium-sized developments, the extensive modelling which is inherent in this stage of the methodology can be safely sidestepped.

The second of the two stages, that of **Building the Database**, again can be treated in a very formal way. The stage involves the application of a set of 'First-Cut Rules' to the logical data model, in order to turn it into a physical database schema. These rules simply enable the analyst to produce a database structure which will 'work' within the system; it will not work efficiently, or even necessarily to a satisfactory standard of performance. At best this structure can be used to support some of the versions of prototypes continuously being developed.

The tuning of the database design to meet the performance criteria set down in the specification is carried out by a designer with detailed knowledge of the particular DBMS software, and so the techniques used must of necessity be outside the scope of any general methodology. However, when the final version of the database schema is produced, revised versions of the Path Analysis Diagrams, illustrating the access to the physical files (as opposed to the logical entity types), may be provided by the database design team, to enable the analysts to bring the final version of the prototype programs into line with the database.

Again this is a quite detailed series of steps, designed to guarantee that a thorough and optimum database is constructed. It is an ideal, even essential, approach when the system being developed is large, complex and critical, but it is clearly excessive when the system is small, and the database design is straightforward. During the Feasibility Study for the project, decisions must be made about the appropriate level of formality to apply to this task.

3.3.5 Detailed Design of the System's Processes

The two stages covered here provide modelling techniques to support the main on-going task of improving the initial prototypes, through iterations and versions, until they are of an acceptable standard for implementation. This is illustrated in Figure 14.8. The prototyping represents the *design* part of the task, while the modelling provides additional levels of *analysis*, helping the development team to identify gaps and weaknesses in the evolving system. Whereas the prototyping of individual system components is carried out separately, by small teams of user and analyst, the building of models such as the Entity Life Histories and the Computer DFDs are done centrally, and mainly by analysts.

The process of **Confirming the Process Detail** involves using the entity types identified during data modelling, and examining each in detail, recording all events which effect it (ie. create, delete or update an entry). This can be a very time-consuming activity, so decisions on whether or not to use it can have a major effect on project costs.

In general, it is always wise to carry out ELHs on the most important of the entity types, usually those which represent the system's main transactions. It may only be necessary to produce lists of event/effects for these, rather than full blown diagrams. On the other hand, when the system being developed is large and the processes are dispersed, or if it is complex in nature, or has critical implications, then a much fuller model-building exercise is called for. The reverse equally applies when the system under development is small and trivial; the process can be handled informally, and no modelling may be required.

The second process, **Applying Necessary Controls**, again depends on the criticality of the proposed system, and how much security and control is required. Much of this can be gleaned from the type of system, and the volume of transactions, the amount of capital, the perceived risk, etc. The process can involve a full study by a team of analysts and auditors, or it can simply be an informal discussion with users on possible danger points. The appropriate level at which to tackle this must be decided upon during the Feasibility Study investigation.

The third process, that of **Assembling the Computer System**, represents the essential activity of pulling the individually developed prototypes together to form suites and sub-systems of programs. This formalises the computer components of the business system being developed, into a Computer System. As part of this, it provides the formal description of the 'physical' new system, in the form of a Computer DFD, which becomes an indispensible part of the system documentation. In some developments, this process may be simple, and may involve very little resource, whereas in others, extensive modelling may be required.

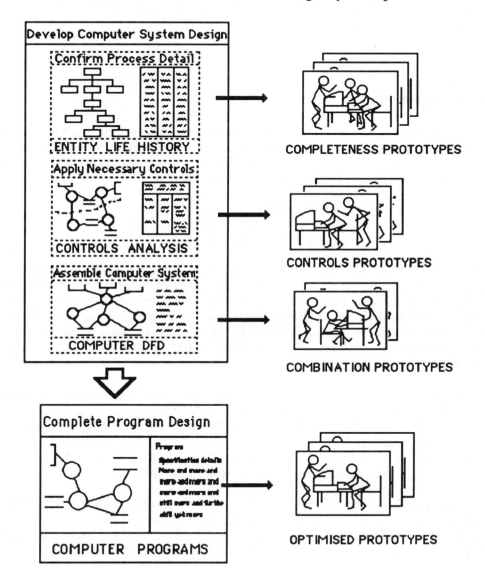

Figure 14.8 Detailed Design of Processes; Tailoring Options

315

The final stage in the methodology involves the completion of the program design, and this includes making use of information about the optimised database structure to review the data accesses within the various computer programs. It also incorporates the acceptance of the finished product by the organisation. The detailed activities in the stage are very hardware/software dependent, and for the most part do not involve the kind of abstract modelling on which much of the rest of the methodology is based. As a result, this examination of the tailoring options within the methodology does not really apply.

4 SUMMARY

This chapter has been concerned with the process of adjusting an evolutionary development methodology to suit the particular development circumstances, and it has suggested broad guidelines as to how these adjustments might be applied at three different levels:

> the organisation
> the fourth generation environment
> the particular project.

The objective of this tailoring approach is to minimise the use of resources, and to optimise the effectiveness of the systems developers. The traditional view of structured systems development has been one of thoroughness and safety, where developers have often gone through the stages and the motions, sometimes having little understanding as to what contribution a particular activity was making towards the final system. This can be very de-motivating.

It is essential that modern systems are developed in a 'lean' and professional way, where the developers know exactly what they are doing, and why it is being done. The gathering of unnecessary information, the duplication of effort, the delay in commencement of stages, the over-formalisation of communications, the excessive documentation, are all part of dangers of applying stuctured techniques in too rigorous a manner.

It must also however be recognised that every short-cut taken during the development of a system carries with it a certain amount of risk. The art of tailoring is one of balancing these risks against the benefits of only applying resources to the areas where they are clearly seen to be needed.

15 PROJECT MANAGEMENT IMPLICATIONS

One of the greatest dangers seen in taking an evolutionary approach to systems development rather than the more traditional pre-specification approach is a potential lack, or even loss, of control during the development process. It can be imagined how easily an unstructured, gradual development, where solutions are simply 'tried out' on users, can degenerate into a seemingly endless series of petty alterations. In such a situation, where no one can be sure when or whether the project has been completed, un-monitored resources are gobbled up without there being any guarantee that the project will eventually prove profitable.

It is ESSENTIAL for the credibility of the whole class of methodologies devoted to evolutionary systems development that an EDM can be shown to respond successfully to a rigorous project management approach.

Project management is the process of planning, organising and controlling the development of a system from its inception through to its delivery and acceptance. This involves the application of various forms of **resource**:

> staff
> hardware and software
> overheads
> time

to some form of development **method**, to achieve a satisfactory level of quality.

The development method to be used must comprise a series of stages, tasks and activities, each of which can be treated as a **measurable unit**. These units can then be used as a basis for the planning and control of resources in the following ways:

> estimation of resources required
> scheduling use of resources
> allocation of work
> measurement of work completion
> monitoring of progress
> adjustment of schedule.

As this book is concerned only with evolutionary development methods, this chapter will examine how the use of such methods affect the way in which project management should be carried out. The following four topics will be explored:

1. The nature of EDM project management

2. Planning an EDM project

3. Monitoring an EDM project

4. Conclusions and implications.

It must be stressed that this is not an attempt to provide a new project management methodology to deal with evolutionary development. The intention here is simply to prove that the evolutionary approach can be managed just as effectively, if not more so, than any traditional 'life-cycle' based methodology.

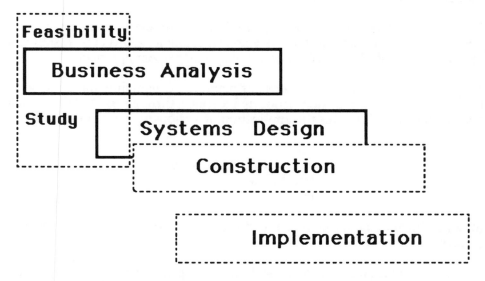

Figure 15.1 The Stages of a Systems Development

1 THE NATURE OF EDM PROJECT MANAGEMENT

It has already been pointed out that one of the most important differences between a standard structured analysis and design methodology and an EDM is that, whereas the product of the analysis and design methodology is a 'specification' which is ready to be coded, the product of an EDM is a fully developed system ready for implementation. This is not to say that an EDM encompasses all aspects of the project development. The EDM relates only to the use of modelling techniques during the analysis and design

318

stages of the development. Prototyping is of course considered to be one of these modelling techniques, and an evolutionary prototype, of its nature, incorporates elements of construction and implementation. However, there are many other components of a project development (eg. interviews, hardware procurement, cost-benefit analysis, development of test and implementation plans, etc.) which are not encompassed within the EDM, but must be considered in the overall planning and management of the project.

As an example of this, Figure 15.1 shows the two stages in the Systemscraft methodology, Business Analysis and Systems Design, within the context of a full systems development. It illustrates the three other stages which overlap and are influenced by the methodology stages. These obviously also require to be managed just as much as the other parts of the project.

Throughout the development process there are three major considerations which differentiate an EDM approach from a more traditional one:

1. The people involved in the development

2. The types of activity that make up the approach

3. The methods of project control

1.1 The People

The basic philosophy of evolutionary development differs from that of the more traditional development approaches in the relationship between the users/owners and the analysts/designers. This relationship is seen as a kind of partnership, where each partner brings some special expertise to bear on the problem. In the case of the user, the expertise relates to the business aspect, and the analyst contributes the necessary technical knowledge and skills. It is true to say that the user's contribution to an EDM project is much greater in terms of time, of cost and of quality, than is likely to be the case for a normal life-cycle development.

This equality of status and commitment to the development is echoed in the fact that the contributions of both the analyst **and the user** must be scheduled, estimated for, costed, and charged to the project. This is very different from the situation in most traditional projects, where the users' time is rarely included in the cost-benefit analysis. This means that most users are given no credit for any contribution they make, and as a result their level of commitment is low. On the other hand, in an EDM development, the individual quality of each of the system components, as well as the ultimate success of the whole project, is seen as the responsibility of the user, just as much as if not more so than that of the analyst.

In several evolutionary methodologies, this relationship is recognised in the structure of the working units of the project team. For example, in Hoskyns' 4GDM, each prototype is handled by a 'development team' which consists of one or two 'user-designers' and one 'analyst-prototyper'. This team is given the task of designing, coding and testing the prototype, then presenting it to representatives of the project management (normally senior users) for final acceptance.

It is also suggested that wherever possible, the project manager should be appointed from the ranks of the user department involved, rather than from the IT personnel. When this approach is taken, a senior IT 'advisor' must also be appointed, and the two must work closely together on matters such as project planning and estimation. The main advantage gained from having a user project manager is that it places a heavy emphasis on the user's continuing 'ownership' of the business system, and stresses the advisory nature of the IT department's role; analysts and designers must begin to see themselves as consultants in the development process, rather than as developers.

The one disadvantage of putting senior users in charge of projects is that once a project is complete, the user-project manager will probably revert to the role previously held, and the build-up of experience obtained through managing that project may never again be utilised. It does have to be recognised that there are unique skills required for managing a systems development project, and that it takes some time to acquire these skills. It would therefore seem sensible to make use of our project managers over a whole series of developments.

1.2   The Types of Activity

A systems development project is made up of a number of stages (see Figure 15.2) and each stage is itself made up of a number of functions or tasks, each of which may be split into a several smaller activities or sub-functions. These represent the **measurable units** mentioned earlier in the chapter. For each of these lowest level activities, plans must be drawn up, estimates of resource requirements must be made, staff must be allocated, progress must be monitored, etc.

It is very important to recognise that although most of the activities may need to be carried out irrespective of whether a standard life-cycle methodology or an EDM approach is taken, their order and sequence is likely to be very different, and the amount of resource required may also vary significantly. A good illustration of this occurs in the 'Implementation' stage of a standard development project. This consists of a very large number of tasks and activities, including for example 'systems testing' and 'user acceptance testing'. However, in an evolutionary development, many of these may be carried out much earlier, perhaps as part of the normal iterative prototyping process.

Many of these individual activities can be classified as being of the same type. For example, 'interviewing' is a type of activity, which may occur in several places throughout the project, during both the Feasibility Study and Business Analysis stages. General standards and guidelines relating to performance and measurement can be given for each of these activity types.

The majority of such activity types are common to all forms of systems development, but two of the most important , those of **modelling** and **prototyping,** have special significance for EDM project management.

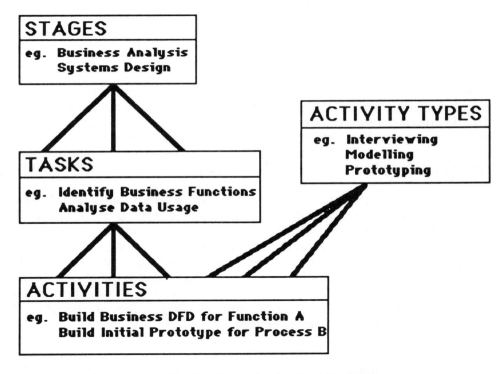

Figure 15.2 Measurable Units for Project Management

1.2.1 Modelling

The structured analysis and design models, such as the Business DFD and the Entity Life History, which are built as part of the project, lend themselves readily to the processes of measurement. It is possible to devise a set of estimation guidelines for each of the different types of model, and it is relatively simple to measure the amount of resource used to create and establish them during the development process. All modelling involves three basic steps:

1. The collection of necessary information

2. The building of the model

3. The testing of the model for accuracy and correctness

This last step is itself a group activity, normally carried out in the form of a structured walkthrough. Problems identified during these walkthroughs may force an iteration of the steps, in that the model may have to be revised, and re-submitted for validation. For each modelling task in the project, estimates can be made for each of these steps, and developers can be required to provide figures of actual time and resources taken.

1.2.2 Prototyping

A popular myth about prototyping is that it is an open-ended technique for establishing user requirements, and that the amount of time involved will depend largely on the whim of the user. In this modern era of cost-justified software engineering, this kind of attitude is unacceptable. In an EDM project it is imperative that the number of prototypes, and the amount of resources required for each, can be estimated for, measured, and controlled.

Firstly, it is necessary to identify the number of different prototypes that are to occur in the project. In some approaches, one large prototype is built representing the whole system, but in the majority of cases the system is divided into smaller components, each of which is prototyped separately. When this happens it is important to formalise the prototype boundaries in order to avoid the danger of duplication, and to facilitate the measurement and control of development progress. Chapter 9 of this book provides a detailed example of how this can be done, illustrating how the problem is tackled in the Systemscraft methodology.

Almost all evolutionary methodologies make use of the concept of prototype **versions**. A version is an issue of a prototype which contains a specific definable level of functionality. An early version for example may contain only the basic dialogue for processing simple transactions, and as the system development progresses, more functionality can be included in later issues. This is now a well-established way of controlling the evolutionary process. Figure 15.3 gives a suggestion of the successive versions which might be created for prototypes in large and in medium-sized developments; (the scope of each of these versions has been discussed earlier in the book).

In the case of a simple development project, often a single version of a prototype will suffice.

Within each prototype version, there are likely to be a number of iterations. An iteration is a release of the prototype for user examination. If the user finds faults, or makes suggestions of new requirements, then the prototype can be adjusted and re-released. Various methodologies suggest limits on the number of iterations that should be allowed for any version of a prototype: the consensus of opinion seems to be three.

EVOLUTIONARY PROTOTYPE VERSIONS	
Large Systems	Medium Systems
Initial Completeness Controls Combination Optimised	Initial Full-Function Optimised

NON-EVOLUTIONARY PROTOTYPES
Pattern-Book Performance

Figure 15.3 Prototypes used within an EDM

So the activity of prototyping comprises a number of sub-activities:

1. Discussion among the 'development team' on the format and content

2. Construction of the prototype using a 4GE: (work carried out by the analyst-prototyper)

3. Exercising the prototype: (a complex task which includes elements of system testing, and carried out by the user-developer)

4. A number of iterations to revise and perfect the prototype version

5. The presentation of the prototype to the project team for acceptance   (normally carried out in the form of a walkthrough).

Again, each of these sub-activities can be estimated for, and the figures can be used to help calculate the potential cost and therefore the feasibility of the project. Later, when the project is under way, the actual work can be measured, and compared with those estimates.

There are two kinds of prototype which may occur within an evolutionary development, but which are not themselves considered to be evolutionary (at least not in the sense in which we have been using the term throughout the book). These are **Pattern Book** and **Performance** prototypes.

A Pattern Book prototype is normally used during the Feasibility Study stage of the project, and its purpose is to make user management aware of the types of system that are available and might be suitable for their organisation. It inherits its name not from its function (as in the case of the other prototypes) but from the manner in which it is made use of: a number of options are demonstrated to the users, who will make their selection based on advice. Such prototypes do not normally need to be built specially for the purpose, they are often simply examples of other similar systems, and are only used to illustrate 'look and feel'.

On the other hand, a Performance prototype, if it is needed, will be carried out as part of the 'Analyse Data Usage' and 'Build Database' tasks within the Systems Design stage. Its purpose is to test the hardware-software capabilities of the proposed system environment, to check that it is physically capable of handling the volume of transaction that the system will generate. This can be one of the most important activities in the whole project. **Obviously, if the proposed system can be proved to be incapable of working, then there is little point in building it!** When there is genuine doubt about the feasibility of the project then a Performance prototype must be built as part of an extended Feasibility Study.

A Performance prototype differs from an evolutionary prototype in two important ways. Firstly, the analysts are not concerned with the content of the transactions being tested against the environment, only with the volume and frequency. As a result, transactions can be 'simulated', and as such do not constitute part of a genuine working model of the system. Secondly, because of this fact, the users are not involved in the construction of the prototype, nor in the analysis of the results. The whole operation must be carried out by experts in database technology.

These non-evolutionary kinds of prototype have been mentioned here because they do not fit into the activity type as discussed earlier. This means that different factors must be used in estimating the resources required.

1.3 Methods of Project Control

It is essential that any approach to project management provides a communications structure, whereby all necessary information is available at the start of each stage or task, and any changes which occur during the execution of the task are notified and acted upon. In the traditional life-cycle approach, information is collected, analysed and formalised into text specifications. Stages and tasks are completed and signed off, and the information produced is recorded in some form of **documentation**. This documentation then becomes the input to the succeeding dependent tasks. Elaborate change-control systems are devised to handle exception situations, where a document which has been accepted as complete is found to require alteration.

One of the main strengths claimed for an EDM approach is that there is a much reduced need for formal documentation. The detailed information about the existing system and the user requirements does not have to be 'captured' and frozen before the analysis and design processes take place, because the 'experts' who have this knowledge at their fingertips are continuously involved in these processes, and the information can be accessed directly from them when needed!

It is not only the communication between analyst and user which is different in an EDM, the use of 4GEs and analyst-programmers make it unnecessary to provide a formal program specification as a communication document between the analyst and the programmer.

The stages and tasks of an EDM are generally similar to those of other methodologies, except that many of them take place much earlier in the cycle, and because a number of them are being carried out **simultaneously**, the full formalised information required for their execution is not available beforehand.

For example, in Systemscraft, the task of building an initial prototype commences when the development team are given a Systems DFD with the prototype boundaries indicated. There may also be provided with brief process descriptions for the computer and user components of the proposed prototype. However, during the actual building and exercising of the prototype, much new information will come to light, it will be analysed by the team, and decisions will be made based on it. Some of this information is likely to be of relevance elsewhere within the project, perhaps to a second development team working on a different prototype. The information may refer to the data structure within the system, and therefore it may affect the construction of the Entity Model, which is perhaps being carried out by a separate team of data analysts. How is this information to be circulated? How is this analysis to be made available to those who need it? How are these decisions to be cross-checked with other parts of the system? There must be **mechanisms** for the capture, communication and transfer of information between the stages and tasks, and between the people involved in their execution.

In almost all EDMs, the main mechanisms for the holding of information relating to the project are the MODEL and the PROTOTYPE, and the communications medium by which their information is dispersed is the WALKTHROUGH.

BUSINESS ANALYSIS	MODELS	
TASKS	TYPE	NUMBER
Identify Business Functions	BFD	1
Identify Business Information Flows	DFD	n
Analyse Business Information	EM	1
Confirm Business Information Structure	RM	1
Incorporate New Requirements	various	n

SYSTEMS DESIGN	MODELS		PROTOTYPES
TASKS	TYPE	NUMBER	NUMBER
Identify Computer System	DFD	n	——
Establish Human/Computer Interface	LPA	n	n
Confirm Process Detail	ELH	0 or n	0 or n
Apply Necessary Controls	various	0 or n	0 or n
Assemble Computer System	DFD	n	n
Analyse Data Usage	NM,DUC	0 or 1, n	——
Build Database	Schema	1	——
Complete Program Design	—	——	1

Figure 15.4 Systemscraft; Stages, Tasks Models and Prototypes

Figure 15.4 gives an example of how in one EDM the products of all stages and tasks are in the form of models and/or prototypes.

1. Each model is built by at least one analyst, and must be checked thoroughly in a walkthrough before being used as

input to another task. The walkthrough is populated by representatives of all groups who have an interest in the area being modelled.

2. Each prototype is built by a development team who will exercise and iterate it until they are satisfied that it is acceptable. The prototype is then subjected to a walkthrough, similar in type to the model walkthroughs, but perhaps with a stronger user-orientation.

The whole series of walkthroughs which make up the quality inspection and control of the project are organised and monitored by a **Co-ordinator**; a senior member of the project team who has responsibility for Quality Control, dissemination of information and maintenance of documentation (including the data dictionary).

The Co-ordinator is responsible for ensuring that all the right people are invited to contribute to and learn from each of the walkthroughs. These include representatives of all groups who create input for or take output from the task of which the model is part. At the end of a successful walkthrough, all members will accept the validity of the model, and will be required to bring their own work into line with the agreed findings.

If for some reason it is necessary to make a CHANGE to a model after it has been accepted in a walkthrough, it may be necessary to re-convene the walkthrough, and address the implications of the change. This again is the responsibility of the Co-ordinator.

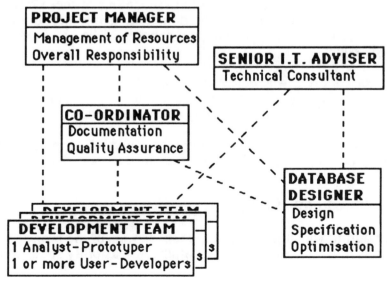

Figure 15.5 The Staffing of an EDM Project

Figure 15.5 illustrates the control structure of an EDM project, highlighting the roles which must be adopted by project staff. It should be emphasised that the names in the diagram represent roles not people. It is quite common for the same person to adopt more than one of these roles, and in fact, in the most extreme situation of a very simple end-user project, the whole development might be carried out by one individual!

At the end of each of the two stages, Business Analysis and System Design, a formal report is produced. The content of these reports has been discussed during parts 3 and 4 of the book, but it is worth re-iterating that most of each report consists of the models produced throughout the stage. These reports, which are directed primarily at the user-owners of the system, should be relatively short, and it should not prove to be difficult for these users to understand them. After all, they or their representatives will already have accepted the models during a formal walkthrough!

2 PLANNING AN EDM PROJECT

The process of planning the development of any form of IT project is extremely demanding and requires great skill. The main difficulty relates to the number of different ingredients which must be considered and weighed against each other. The process involves several major activites, which in theory should be done in sequence, but in practice almost always require to be done simultaneously. These include the initial examination or 'scoping' of the project, the identification of the resources required, the consideration of the availability of resources, and the scheduling of these resources to carry out the work. All this must be done taking into account the costs involved, and balancing them with the benefits to be obtained from the completed system.

Planning does not stop when these initial schedules are drawn up. Once the project is under development there is a continuing requirement to revise the schedule in light of the unfolding reality of the situation.

Here, the aspects of these planning activities which relate specifically to an evolutionary development approach are examined in more detail.

2.1 The Feasibility Study

The stage during which the most important aspects of the project planning take place is of course the Feasibility Study. A feasibility study comprises three tasks:

1. An initial investigation: (the content of this has already been described in detail in the previous chapter).

2. An analysis of the findings: this involves the high level planning, scheduling and costing of the proposed development process.

3. A cost-benefit analysis, where an estimate of the usage of resource, both in terms of cost and time, are compared with the expected benefits from the new system.

The feasibility study is formally completed with the presentation of recommendations to the user-owners of the proposed system.

In the case of a project for which an EDM type of development is anticipated, these tasks would be carried out very quickly; the normal duration of such a study would be between one and three days. In some circumstances however, the initial findings may suggest the need for a more detailed examination before the feasibility can be fully established.

It is impossible to over-stress the importance of this initial investigation and analysis. The eventual success or failure of the project is often judged on these early estimates; after all, these are the same estimates on which the decision to proceed was based! For this reason, feasibility studies should generally be conducted by senior, experienced analysts, with proven ability in sizing projects.

There should also be some continuity between the conduct of the feasibility study and the main development of the system. It is suggested that one of the analysts responsible for the study should be appointed to the full project, perhaps as senior IT advisor. In this way, some form of accountability can be established for feasibility estimates and recommendations.

An alternative view can also be put forward, suggesting that these early estimates are better made by someone who does not have a personal interest in the development, and is therefore more able to be impartial. Both points of view are valid, and both can met by the appointment of a two-person team to carry out the study.

2.2   Planning and Scheduling

So the project plan is drawn up during the feasibility study, after the initial investigation has taken place. There may of course be important aspects of the project which are identified at this stage, and must be included in the overall plan, but which are outside the scope of an evolutionary development methodology. For example, there may be a need for **hardware procurement**, and/or for the **development of a data communications network**. There are various techniques for the development and control of these, and the reader is advised to search these out as and when they are needed: here we are restricting our examination to the area of the analysis, design and construction of information systems software.

Planning and scheduling involve examining each of the project tasks and activities in detail, and making decisions on the order and sequence in which they will be carried out. This is best illustrated using a specific EDM as an example. As before, Systemscraft is used, but the principles expressed during the discussion apply equally to other EDMs.

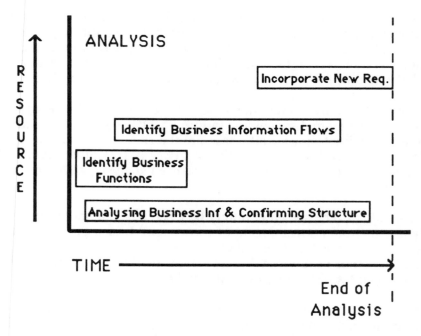

Figure 15.6 Systemscraft; Schedule for the Analysis Stage

2.2.1 The Analysis Tasks

According to the methodology, there are five basic tasks which constitute the Business Analysis stage of development. These are:

> Identify Business Functions
> Identify Business Information Flows
> Analyse Business Information
> Confirm Business Information Structure

330

Incorporate New Requirements.

These tasks involve the production of four major models that together make up the complete Business Model. Initial versions of these will have been constructed during the feasibility study, so this more detailed study will hopefully confirm, improve and expand them, and will provide all necessary supporting documentation.

Figure 15.6 illustrates how these tasks may be scheduled in terms of time and resource usage. This is done in the form of a simplified Gantt chart, where activities are allocated start times based on inter-task dependencies, and on the availability of resources (usually staff). The following points give the reasoning behind this particular example.

1. The first task, Identify Business Functions, involves the construction of a hierarchical Business Function Diagram, on which much of the rest of the business analysis is based. Where the business requirements are standard, and already well-understood, this should take a relatively short time. However, where requirements are vague, and management are undecided about the project scope, much more time and resource must be scheduled. In all cases, the BFD model must be well-advanced (and ideally accepted) before the task of Identifying Business Information Flows begins.

2. This second task involves the building of detailed Business Data Flow Diagrams. Although it is shown here as one activity, it is quite usual to split the task, allocating different parts (or functions) to different analysts. Often the building of DFDs for various parts of the system is staggered, to make best use of resources. (This is shown more clearly in Figure 15.10, where work on the three main functions of a business system are scheduled separately.)

3. The two data-oriented tasks, which provide the entity and relational models, are combined in Figure 15.6, and treated as one long activity for scheduling. The two tasks always overlap, and in a small system they can prove difficult to separate. However, in the case of a sizable development, the combination of the two tasks would represent too large a unit for control, and it would therefore be better to formalise and schedule them as separate.

4. The diagram shows the Incorporation of New Requirements as a separate task on the schedule. As was suggested in the previous chapter, there is often a good case for incorporating this task as part of the other four tasks.

The 'end of the analysis' shown in the diagram represents the official signing-off of the Business Analysis stage, where the Requirements Specification is formally presented to user management.

2.2.2 The Early Design Tasks

Two important design tasks overlap considerably with the analysis stage. These tasks are:

>   Identify Computer System
>   Establish Human/Computer Interface.

Both tasks must be considered as part of the design stage, because they involve decisions on the physical form of the new system. But equally, the discussions carried out with users during these tasks provide important material for analysis, and as such, supply inputs to the analysis tasks described earlier. Figure 15.7 illustrates the extent of this overlap.

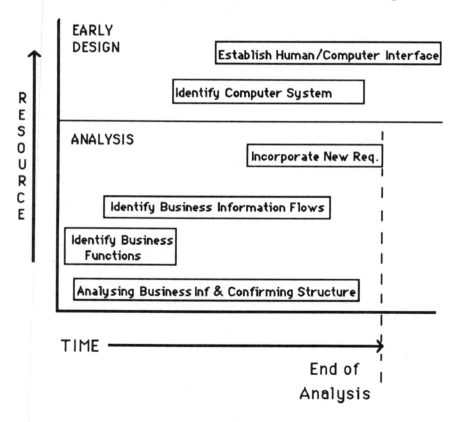

Figure 15.7 Schedule for Early Design

There are two important points to be made concerning the scheduling of these early design tasks;

1. Both tasks have the same property as the earlier task of Identifying Business Information Flows, in that different functional parts of the system can be scheduled separately and allocated to different staff: (again, Figure 15.10 illustrates this).

2. The second of these tasks, Establishing Human/Computer Interface, involves the construction of prototypes. This means that the earlier this task is scheduled, the sooner the users can begin to take an active part in the design process. Ideally, this should happen within days or weeks of the start of the project.

One possible problem with this overlap of analysis and design tasks is that most of the existing project management methodologies do not recognise that such an overlap will occur. It is common practice for organisations to treat the formal end of the analysis stage as a potential 'cut-off' point, when work on the project can be abandoned, or the later stages postponed. In an EDM approach, if the project is abandoned at this stage, greater overheads will have been incurred, and it is difficult to postpone later stages because the traditional 'integrity' of a life-cycle stage has not been maintained. Nevertheless, if these problems are recognised, they can easily be circumvented.

2.2.3 The 'Incremental' Approach to Scheduling

So, once the very first model, the Business Function Diagram has been built, the project planners are able to see the scope of the development, and can make judgements on the amount of work involved.

Although some of the sub-tasks and activities are dependent on others, and cannot be started until these are completed, many components from the same stage can, in theory, be carried out simultaneously. It is often not possible, however, to work on all these components in parallel, as this would require too many staff. It is therefore necessary for the planners to decide which aspects of the business system should be given priority, and to schedule the order in which the different parts of the development should be carried out.

Traditionally, the order in which such components were developed could be based simply on the convenience of the development team; it did not really matter what was completed first, because **all the components had to be**

completed before the system could be implemented. There were always of course some opportunities for 'phased' development, where the project could be split into a series of integral sub-projects, the development of which could be carried out and implemented one after the other in an incremental fashion.

Phased development is itself comparatively uncommon, and in any case does not match up to the evolutionary idea of organising **every possible functional component of the system to be released as soon as possible.** This approach is dealt with in much greater detail in Tom Gilb's book *Principles of Software Engineering Management.* The rationale behind it is that some parts of the system can be developed and implemented quickly, while other parts are still being designed. So, from an early stage of the project, these completed parts are providing user benefits, and in some way are helping to pay for the continuing development. In this way, not only do we get 'evolutionary development', but also 'evolutionary delivery.'

But how is this to be done? What criteria are to be used to decide on the sequence and priorities? Here are some of the kinds of deliverable component that the planners may wish to look for when scheduling the sequence of development and release:

1. Naturally integral units; parts of the system which function with minimal connection to other parts. (The identification of these should be aided by the 'functional' modelling approach taken during the early investigation)

2. Smaller units to be done first, to set in motion the flow of system benefits as early as possible. (Sometimes these units may not be physically small in terms of number of instructions, but may involve only a small amount of development work: for example, a unit may be built by 'modifying' another 4GE-built system with similar attributes)

3. Parts of the system which provide the most 'payback' to the user for the least amount of development work

4. Highest profile parts of the system, most likely to impress the users

5. Parts of the system where the users are most supportive and enthusiastic; (good progress on these should help to 'prove' the approach to the less committed users)

6. Parts of the system to be provided in the natural sequence in which they normally take place; (eg. the validation of orders before the allocation of stock). This may enable products from the early prototypes to be used as support for the development of the later parts.

The planners should also bear in mind that the provision of small but regular deliveries is most effective, so that the users are continuously obtaining the rewards for their investment, and the changes to the system can be seen as a gradual series of improvements rather than as one major 'upheaval'. **In other words the new system should 'seem' to be evolving from the old, even though it is the result of a complete new analysis and design.**

It is important to remember however, that there are still many systems for which it would be unwise to take an incremental delivery approach; a one-off, full and proper changeover from the old to the new system must be organised and co-ordinated, often after a period of parallel running. Nevertheless, even for these systems it will still be necessary to schedule the sequence in which the early design processes are to occur.

2.2.4 The Detailed Design Tasks

The rest of the Systems Design stage involves the detailed evolutionary development of the design for both the data and processes of the system. The tasks, according to the Systemscraft methodology, are:

> Confirm Process Detail
> Apply Necessary Controls
> Assemble Computer System
> Analyse Data Usage
> Build Database

It has been pointed out several times earlier in the book that if the system to be developed is small, straightforward and non-critical, a number of these tasks may require only a cursory examination, and it may not be necessary to produce the formal models and prototype versions associated with them.

Figure 15.8 does however illustrate the situation where all tasks are required, and the following points arise from that illustration;

1. It should be noted that the **modelling** activities in the three process-oriented tasks differ from those of earlier processing tasks in that they involve examining the system **as a whole**, rather than dividing it into a series of integral functions. This means that some kind of standard definition

of the complete system must be available as input to these tasks, and it suggests that **they can not start until the formal Business Requirements Specification has been issued** (at the end of the analysis stage). This is a very important dependency from the point of view of the planner and scheduler, and it is highlighted in the diagram below.

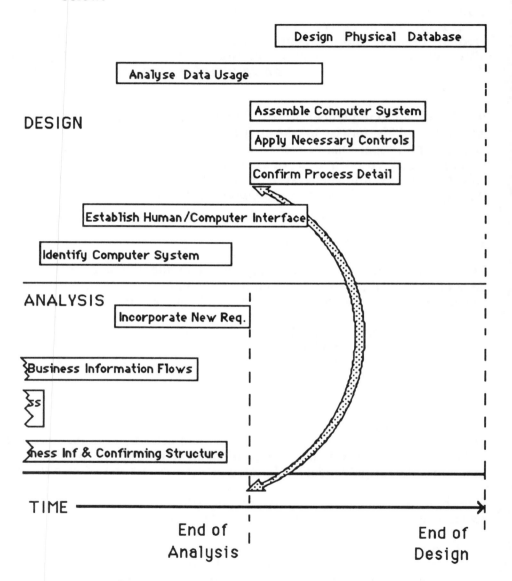

Figure 15.8 Systemscraft; Schedule of Detailed Design Tasks

2. On the other hand, these same three tasks also involve the construction of new versions of the earlier initial prototypes to take account of the findings from the structured modelling activities. There is a strong case for scheduling these **prototyping** activities separately, particularly as they are likely to be carried out by the same development teams responsible for the initial versions, whereas the modelling activities will be conducted centrally.

3. A further dependency should be noted concerning these three tasks. It has been suggested that each might result in the production of a new version of the prototypes:

Confirm Process Detail	*Completeness*
Apply Necessary Controls	*Controls*
Assemble Computer System	*Combination.*

   It must be obvious from the way in which prototypes are evolved that it is not sensible to have more than one version of a particular prototype released at the same time. As a result, the release of the *controls* version may have to be delayed until the *completeness* version has been fully accepted. This needs to be taken into account by the planners and schedulers. (Again this dependency is best illustrated in Figure 15.10.)

4. The two data-oriented tasks are scheduled here to complete the design stage. The assumption here is that the various versions of the prototypes have been developed using part-databases, which eventually need to be pulled together and optimised. When this is completed, the prototypes can be adjusted, and the final versions can be released (as part of the construction stage). However, there are a significant number of 4GEs in which it is ESSENTIAL that the full database definition be completed before the code for the system processes is constructed. Where this is the case, all the design tasks (and some of the analysis tasks) must be scheduled differently. Some EDMs which are designed purely for use with a relational DBMS have this facet built into their structure.

2.2.5 The Construction Tasks

The Construction stage includes a number of new tasks which are either beyond, or on the periphery of the territory covered by an evolutionary development methodology. However, the main bulk of construction is carried out in the building of prototypes, as part of several of the main **design stage** tasks;

> Establish Human/Computer Interface
> Confirm Process Detail
> Apply Necessary Controls
> Assemble Computer System.

This fact is illustrated in Figure 15.9, where these same design tasks are also shown (as dotted boxes) as part of the 'Construction' stage. The names of the associated prototype versions are given in the boxes.

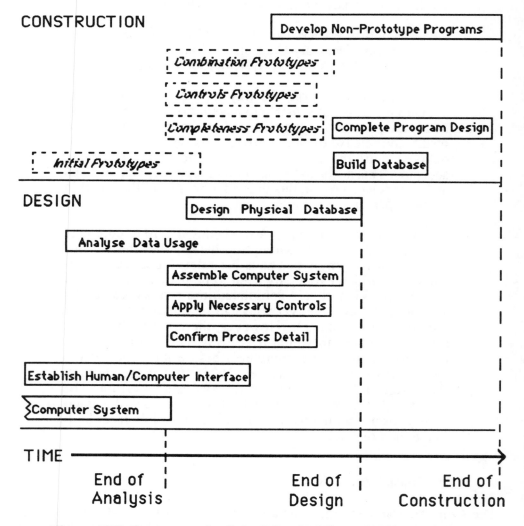

Figure 15.9 Systemscraft; Schedule of Design and Construction

The new tasks involve the physical construction of the database, and the adjustment of the process code to incorporate the final tuning of the database structure.

It is quite common for a system which is being developed using an EDM to have some components which are constructed in a non-evolutionary manner. For example, some areas of the system may require very high performance and may have to be written in a 3GL, and other parts may be designed for batch processing. Another possible activity might involve tailoring systems which are already part of the applications portfolio to interface with the new system. A task to cover such activities is included in the chart shown in Figure 15.9.

2.2.6 Implementation

It has already been suggested that, wherever possible, an 'incremental delivery' approach should be used. There will however always be some systems (perhaps even the majority) which require the more traditional type of implementation. This might include the set-up of master files, a planned changeover (either direct, parallel or pilot), and a formal user acceptance of the system. As these tasks apply equally to non-evolutionary development projects, it is not considered necessary to describe them here. It should be noted however that a number of the entries that one finds in the traditional list of implementation tasks do not apply in the case of an evolutionary development, because they are implicitly included in earlier design tasks.

2.3 Estimating and Costing

There are many aspects of costing and estimation which apply specifically to, or which have particular significance when considering, evolutionary development. Many of these have been touched on earlier in the book, and some, because of the relative newness of the topic, may not yet have come to light.

The main unit which has been used in the past to help gage the amount of resource required for a system development has been the **program instruction**. Planners were able to forecast the likely number of Cobol or PL1 instructions that would appear in the final version of their programs, and then use statistical techniques to help predict costs and resource demand. A well-established example of such an approach is Boehm's COCOMO (Constructive Cost Model) method. This kind of approach is clearly useful if the development method follows the standard life-cycle and is geared to the production of 3GL code. There is of course always the problem of estimating during the feasibility study how many program instructions will occur in the final system!

However, a measurement unit of this type is much less effective within an evolutionary development. Because 4GL code is 'non-procedural', and because different 4GEs vary quite strikingly in their approach, it is not possible to equate a program instruction with the system complexity or resources required for such development. What we need are units of measurement which are language-independent, and which make use of information which can be identified easily during the feasibility study.

One important modern method makes use of such a unit, referred to as a **function point**. Function Point Analysis is a method for determining the relative size of a system based on some measure of the information going into and out from the system, and on the complexity involved in the process of changing that information. FPA involves identifying and classifying the 'system components' (inputs, outputs, internal files, etc), and allocating a number of points to each depending on its type and complexity. The method incorporates the use of several detailed formulae and weighting techniques, and is designed specifically to assist in dealing with the larger systems development.

As has already been stressed, the basic components in an EDM development are models and prototypes. Information on the number and complexity of these should be available after the initial investigation in the feasibility study, and both can be used effectively for estimation and monitoring purposes.

2.3.1   Models

DeMarco, in his book *Controlling Software Projects*, makes use of a revised and simplified form of function point analysis, based on the use of components which can be easily identified from DFD and Entity models. The unit that he makes use of is known as a 'Bang', and is described as an implementation-independent measure of functional requirement. This means that it relates to the complexity of the system requirement, **not** to the task of developing that system, so each different development approach will require additional estimating techniques to associate these units with the amount of effort needed to deliver them. Systems developers are advised to consult this book, and give consideration to the ideas put forward in it.

On a simpler level, planners making estimates may prefer to use a set of *predicted* values of the resource needed for building each of the different types of model in the methodology, then identify how many of each type of model is required to be built. In this way, a figure of resource requirement for the whole development could be arrived at.

Here is an example of the kind of figures that might be used for estimating how long it would take to build the Business Data Flow Diagram. The figures are a predicted average for each separate page of the DFD, and they are based on an expected person-day work load for each of the three model-development steps mentioned earlier:

1. Information Collection   0.5 person-days

2. Model Building   0.5 person-days

3. Testing and Walkthrough   1   person-day

An additional weighting factor may also be introduced, based on the analyst's view of the complexity (or otherwise) of the model.

2.3.2 Prototypes

It must be acknowledged that the time involved in the development of different prototypes may vary considerably. One can not tell beforehand how many iterations a particular version may require, or how long it will take for the user to 'turn it round' (bearing in mind that users will almost always be working part-time on the project).

To allow for such variations a 'statistical' approach must be taken to the task of estimation. Although one cannot be specific about any one particular prototype, it is possible to work out average times taken for the development of prototypes from previously completed projects, and use these as estimate guidelines for new projects. It may be necessary to introduce some form of weighting to adjust for those prototypes which are anticipated to be unusually difficult, or to make allowance for the limited availability of a particular user-developer.

Statistics might for example provide the analyst with the following set of figures for an average individual prototype; (each version of the same prototype is considered for these purposes as a separate prototype). The figures are based on the person-day workload for each of the sub-activities identified earlier:

1. Team Discussion   1 person-day (0.5 user, 0.5 analyst)

2. Construction   2 person-days (analyst)

3. Exercising   2 person-days (user)

4. Iteration   Average 2.5 iterations, each taking
                              1 person-day (0.5 user, 0.5 analyst)

5. Presentation   1 person-day (staff at walkthrough)

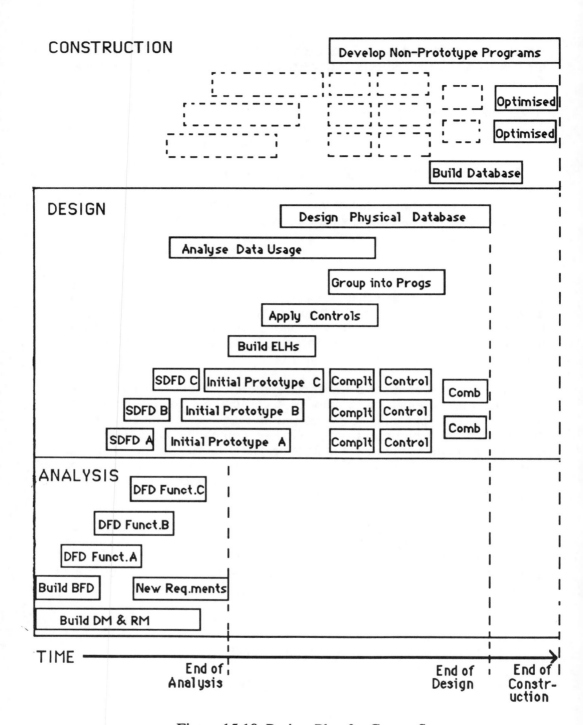

Figure 15.10 Project Plan for Gentry System

This would give an average figure of 9.5 person-days (or 2 person-weeks) per prototype. Again some kind of adjustment may need to be included based on the analyst's view of the complexity of a particular prototype.

It should be emphasised that the figures for both models and prototypes are shown here simply as examples of how calculations can be carried out. They may not be representative of the reader's own organisational methods and practice. **This does not matter!** Each organisation must build up its own statistics.

2.4 Example of Project Plan

The Gentry Order Processing, Stock Control and Purchasing system, which was used as a case study during parts 2 and 3 of the book to illustrate the use of the modelling techniques, can also be used to demonstrate briefly how an EDM project can be properly planned and scheduled.

Figure 15.10 gives an example of a project planning chart for the Gentry system. It does not however contain the full information from the scheduling operation, so some supporting text is included here. The following points should help to clarify the model.

1. The scheduler has decided to use three separate development teams, one for each of the three basic business functions:

    | Order Processing | *Function A* |
    | Stock Control | *Function B* |
    | Purchasing | *Function C* |

    These teams are given responsibility for developing the processes for these function areas, through the building of Business and Systems DFD models on into the construction of several prototype versions. There are actually several different prototypes to be developed within each of the three areas, so a more detailed level of scheduling could have been used if it was felt to be necessary.

2. A separate team is allocated the series of activities concerning data analysis and design. Its work will include the following tasks:

    Analysing Business Information
    Confirming Business Information Structure
    Analysing Data Usage

Designing Physical Database
Building Database

The scheduled start of each of these tasks is staggered, to allow for their inter-dependency, and also for the limited availability of manpower; (see Figure 15.10).

3. Three of the detailed design tasks involve examining the whole system rather than the separate functional parts. These tasks are:

Confirming the Process Detail
Applying Necessary Controls
Assembling the Computer System

Having said this, the models produced from these tasks are distributed to the development teams who can then build the associated prototype versions. This time-dependency is shown on the chart.

4. The decision has been taken to implement this as a full system, rather than as a series of incremental deliveries. Although the implementation schedule is not shown on the chart, it is important to appreciate that the stage must be scheduled, and resources must be allocated.

Although this example is deliberately sketchy, and some vital information is missing, I think it can be seen to provide enough evidence to suggest that a medium-to-large evolutionary development project can be planned to an acceptable level of depth.

3 MONITORING AN EDM PROJECT

Once a development project is under way, it is essential that details are kept of the amount of work completed and the resources used. The units of work can be recognised and recorded as complete when they are 'signed-off' at a walkthrough, but how can we tell how much project resource they have consumed? Equally, how can we tell whether the actual resources used were greater than those estimated?

The most important type of resource is staff-time, and this is recorded throughout the project by means of a **time-sheet**. Figure 15.11 gives an example of a time-sheet which has been designed specifically for an EDM type of project.

TIME SHEET

Project Id.

Name _ _ _ _ _ _ _ _ _

Staff Id.

OTHER ACTIVITY	CODE	HOURS

Week No.

MODEL ID.					
TYPE					
ACTIVITY	HOURS				
Collection of Inf.					
Building model					
Walkthrough					

PROTOTYPE ID.					
TYPE					
ACTIVITY	HOURS				
Pre-build Discussion					
Construction					
Exercising					
Iteration					
Walkthrough					

Figure 15.11   Time Sheet for an EDM Project

It is imperative that **the users** who are involved with the project also record the time they spend working on the different project components. If this is not done, the figures produced for project resource usage will be seriously inaccurate.

In order for a time-recording system to work effectively, all the stages, tasks and activities, and their related models and prototypes, must be uniquely identifiable. This means that some form of **coding system** must be used. Names or codes must be allocated to all the project steps and components during the Feasibility Study, at the time when original estimates are made. Staff can then use these same codes to record their activity.

This coding system must be kept as simple as possible: if it is made too complex, it will be ignored or mis-used. One very useful way of simplifying the completion of time-records is to computerise the time-recording system. In such a system, 'help' facilities can be provided, and the resource usage information is captured directly, making it easy for project managers to evaluate its significance. Evidence shows that computerised time-recording systems are much preferred by project workers, and that more accurate records ensue.

Process Mgmt & Stock Control	PROJECT CONTROL CHART						
PROTOTYPE NAME				P1	P2	P3	P4	P5
1. **INITIAL**	Date Issued	Est.	20-5	27-5	30-5	1-6	6-6
		Act.					
	No. of Iterations	Est.	2	3	2	3	2
		Act.					
	Date Accepted	Est.	25-5	2-6	5-6	6-6	1
		Act.					
2. **FULL FUNCTION**	Date Issued	Est.	1-6				15-5
		Act.					
	No. of Iterations	Est.	2				3
		Act.					
	Date Accepted	Est.	10-6				25-6
		Act.					
3. **OPTIMISED**	Date Issued	Est.	24-7	24-7	28-7	24-7	
		Act.					
	No. of Iterations	Est.	2	2	3	2	
		Act.					
	Date Accepted	Est.	1-8	1-8	5-8	1-8	
		Act.					

Figure 15.12 Example of Document for Monitoring Progress

During our earlier examination of the planning process, a simplified form of Gantt chart was used to illustrate how a project could be scheduled in terms of estimated use of time and other resources. Each project activity was shown as a bar, with its length representing the **estimated** time from start to finish. One of the most effective techniques for monitoring project progress is to record the **actual** time taken for each activity as a separate bar just below the estimate bar. This gives an instant comparison of the estimates and the actuals, and it enables the project manager to gage the state of the project at a glance. Ideally, the chart should also be computer-produced, so that regular adjustments can be made with ease to the schedule and estimates. It may of course be necessary to keep more detailed records for comparison of actuals and estimates. Figure 15.12 gives an example of how this might be done.

Although there are fewer inter-task dependencies in an evolutionary development than in a pre-specification project, there are still major benefits to be gained from the use of 'Network Analysis' to help in monitoring and re-scheduling project work. Network Analysis is also known as Critical Path Analysis, or PERT (Programme Evaluation Review Technique), and is generally considered to be an indispensible aid to the management of larger projects, irrespective of which development paradigm is used. Because of its general applicability, it is not discussed in detail in this book, but readers are advised to consult other sources for more information; not least because network analysis is strongly supported by the main project management software packages, which integrate it with a number of other planning and scheduling techniques.

Finally, when an 'incremental delivery' approach is being taken to the project, there may be some benefit in the use of some form of cash-flow modelling technique, to show how the project is contributing to its own costs!

4   CONCLUSIONS AND IMPLICATIONS

There are two major points which stem from this discussion on the management of EDM projects. The first of these is the importance of maintaining figures on the time taken and resources used for each of the development stages and tasks. This process of collecting what DeMarco calls the 'metrics' of the project has to be part of the culture of the organisation. Often it is neglected by the analyst/ programmers, because it itself takes time, and there is no apparent direct benefit to the system. It is imperative that this collection process is simplified, formalised, and (where possible) automated, and that members of the project team are taught to realise how these figures contribute to the gradual improvement in project estimation standards.

The second point relates to the importance of the use of a methodology **Framework** to the development of the project. Not only does it provide the structure for the methodology components such as models and techniques, but it is also the base on which all project plans, estimates and measures of completion are constructed. There must be a framework, it must be sound and strong, and it must be **relevant** to the particular project being developed. This is why flexibility and scalability are so important in a methodology. It also means that when a 4GE evolutionary development approach is being used, the project management method used must not be based on the traditional 'life-cycle' structure. It may sound like a truism, but surprisingly, it is not always accepted that *the framework of the development methodology and for the management of the project must be the same.*

Having said that, although there are many important differences between the management of an evolutionary and a standard life-cycle project, it must be stressed that most of the existing well-established project management methods can be used in conjunction with an EDM, provided that some aspects are 'tailored' to suit the new approach.

This has been one of the longest chapters in the book. Its purpose has been to prove that evolutionary development using 4GLs and prototypes can be effectively project managed, and therefore, that **it is possible to develop large and complex systems using an EDM**. This is an important step in the establishment of evolutionary systems development as a general purpose method.

It does not mean however that very strict project management standards must be applied to all evolutionary developments. As before, small and medium-sized non-critical projects, which currently make up the majority of evolutionary developments can be managed using a looser form of control than has been specified here. Again, balance and flexibility are the watchwords.

16 SUMMARY AND CONCLUSIONS

The contents of this book represent an examination in some detail of the whole field of evolutionary systems development as it applies to business information systems. It illustrates how the quick and user-friendly methods of fourth generation environment prototyping can be used in conjunction with the more rigorous techniques of structured systems analysis and design. The purpose of arranging such a 'marriage' is to attempt to produce an offspring which inherits the well-proven qualities and benefits of both approaches, and yet suffers from none of the equally well-proven weaknesses. The progeny of this union are what we have been calling Evolutionary Development Methodologies.

One particular example of an EDM, Systemscraft, has been used extensively throughout the book. This methodology is currently being used within a number of organisations, and is available to be made use of by any DP department wishing to introduce such an approach. However, the main reason for putting it forward here was to provide an example 'environment' where the ideas and principles of evolutionary systems development could be examined, and conclusions drawn which could be seen to apply to the whole generic field.

1 ESSENTIAL ASPECTS OF AN EVOLUTIONARY METHOD

Most of these conclusions have been discussed in one form or other during the preceding chapters, but our intention here is to outline the most important of them as a set of high-level guidelines for the development or tailoring of any evolutionary prototyping methodology. The suggestion is that these seven aspects must be present in any such methodology if the full benefits of structured systems analysis & design and prototyping are to be achieved, and that their combined descriptions can be looked upon as an extended definition of an EDM.

1.1 Overlap of the Analysis,   Design and Construction   Stages

One of the most important ideas in the evolutionary development concept is that prototypes must be delivered at a very early stage to the user. In the definition of prototyping examined in chapter 1, it was stressed that the purpose of a prototype was to 'help identify the user requirements', and this identification of requirements is part of the analysis stage of the development. However, in many structured systems methodologies, the analysis stage involves identifying the complete 'logical' requirements before considering how they are to be physically implemented. This means that no prototype (which is by definition a 'physical' model) can be built until the start of the formal design stage. If analysis must be complete before design starts, then clearly the prototyping technique can not be used to assist in the analysis.

In fact, a good prototype should not only check the validity and completeness of the analysis carried out earlier, but can also identify new logical requirements and implications.

Obviously some logical analysis must be done before the prototype can be built; it is after all a physical implementation of the logical requirements! The argument is that the revisions to the prototype are likely to affect both the physical and logical elements, so if the full logical specification has been formalised and agreed, then some of the benefits of using a prototype are wasted.

The approach suggested in this book involves the use of a hierarchical Business Function analysis to identify separate functional components of the system. This can enable the staggered development of the further analysis models, allowing one functional area of the system to be progressed into the design stage while others are still under analysis. This means that the first prototype of a small part of the system can be with the user often within a few days of the start of the systems analysis.

1.2   Limited Modelling of the Existing System

A large number of structured systems methodologies incorporate a stage for the modelling of the existing physical system (usually in data flow diagram form). This can be quite popular with the users, who can recognise their current working practices, and can clarify any misunderstandings in the detail. However, it is often questionable as to how valuable much of this information is. For example, if the new system is to be very different from the old, then many of the existing processes will be completely changed, rendering their detailed analysis almost worthless. (There is of course still a need to create a 'logical' model of the existing system.)

The argument in favour of retaining the modelling of the existing physical system is that the analyst can use it as the starting point for the analysis of the 'logical' requirements. There is clearly some justification in this, in that many of the user's requirements are likely to be implicit in the existing system, and some may easily be overlooked should that system be ignored. However, the overhead involved in finding these few albeit important oversights can be excessive, and the suggestion is that using an evolutionary development approach, these facts can be obtained in another way.

The assumed necessity of modelling the existing physical system is part of the traditional view of the systems analysis process, where the analyst is required to identify and record everything that the user knows about the system, then use that information for the analysis. It has long been realised that it is almost impossible to glean every piece of information about how a user carries out a particular task; some of that knowledge is the result of the combined experience of twenty years, and it would be arrogance to assume that an analyst could pick this up in a half-hour interview!

The traditional view suggests that the analyst transfers all knowledge about the existing system from the mind of the user into a specification to be used by the development team. The team then continue with the analysis and design, and record the completed design in a specification for transfer to the mind of the programmer. This transfer of information involves a massive overhead in specification, and opens up all kinds of opportunity for misunderstanding and error.

In the case of an evolutionary development approach there is no need for this complete transfer; the prototyper and the user are working in partnership on the analysis and design, so that the repository of detailed knowledge about the existing system can remain in the mind (and the documentation) of the user. When these details are needed they can be summoned at will, and only those pieces of information that are relevant need be put forward for examination.

It is easy to overstate this point of view; there must obviously still be 'specifications' produced as part of the analysis and design, but they require to be much less detailed, and their purpose is to formalise the decisions and agreements on the scope and direction of the project. Nevertheless, if the detailed specialist knowledge of the user is not put to effective use throughout the analysis and design stages of the project, then many of the benefits of an evolutionary approach will be lost.

1.3   Partnership and User Responsibility

This idea of the analysis and design of the system being carried out in partnership between user and analyst, with each bringing to the task specialist knowledge and expertise, is an axiom of the evolutionary development creed. The approach has been formalised in a number of different ways, one of the most important being in the writings of Milton Jenkins.

This much greater level of involvement for the user in the development process must be accompanied by an acceptance of some of the responsibility for its progress. The role of the user, for example in the testing and reporting back on prototype iterations, must be formalised, standards must be devised, and the adherence to rules and deadlines must be made to apply equally to both partners.

This inevitably means that the contribution required from the user must be planned for, estimated, costed and monitored in the same way as any other project resource. In the past, it has been common practice not to include user activity as a component in the development cost analysis or in the resource planning; in an evolutionary development such an approach would be unrealistic. As a result it may appear that the resource requirements for an EDM project are greater than those for a more traditional structured systems project. However, from many of the points already made, it can be seen that the opposite is likely to be true in most cases.

The major advantage of this extra user involvement during the development process is the very high level of commitment that it instils. Because the user is 'responsible' for the design, the system is much more likely to be implemented and accepted wholeheartedly. Again, this is a benefit that should spring from a well-structured evolutionary development approach, and failure to establish the user partnership concept could place it in jeopardy.

1.4   Formalising Prototype Boundaries

It is important that when a piece of the proposed system is identified as a unit of work to be prototyped, it should be marked and documented in terms of its scope and boundary. In an uncontrolled prototyping environment there are great dangers of duplication of effort and of uneven coverage of the system requirements.

For example, during the iteration of a particular prototype new ideas may surface, and before deciding to implement them the prototyper must be sure that they are not being incorporated elsewhere in the development. In order to make the decision, each prototyper must have a very clear idea of the functions required from the unit being worked on, and also must have an appreciation of the rest of the system, in particular the parts of the system that interface with this unit. In a well-stuctured development methodology the availability of these two essential areas of information would be guaranteed by the methodology framework.

Most of the methodologies examined seemed to provide little to support this requirement. In the Systemscraft methodology, the problem is tackled using a version of the data flow diagram which separates the human computer parts of the proposed system. Candidate prototypes are identified from this DFD, and marked accordingly. The outline of each prototype is supported by the Process Descriptions of all the DFD processes involved, and similarly by the data definitions of the entity types mentioned within the prototype boundary. The analyst may also have details of early interviews conducted with the user, and is able to supplement these with further interviews before building the prototype. During the building process the analyst may be further constrained by the organisation's standards on dialogue design.

The analyst will also be in constant contact with the project manager, and will be required to be involved in the walkthroughs for all other related parts of the system.

1.5 Evolving Levels of Functionality

For a project to be managed successfully, control must be exercised over the stages within the project, and the progress of each stage and sub-stage must be carefully monitored. In the standard structured methodology approach, these stages are based firmly on those defined in the Systems Development Life Cycle, and the fact that the analysis and design stages of an evolutionary development project overlap can add to the complexity of the control problem.

The use of evolutionary prototyping techniques can also present problems in terms of estimation and monitoring: as yet, analysts have only limited experience of the use of these methods, and this is bound to be reflected in their forecasts and estimates. Also, so much depends on the understanding, the abilities and the attitude of the user. In theory, such a prototype can be set up quite early during the analysis stage, and still be undergoing revision by iteration almost until the release date. As suggested earlier, the fact that a prototype involves some investigation, some analysis, some design and some implementation, means that a form of concept other than a life-cycle stage must be used to split the prototype into measurable and therefore controllable components.

One approach put forward is to make use of the concept of a 'level of functionality' within each prototype. Not only are there ITERATIONS, where corrections and user's suggestions may be incorporated, but there are VERSIONS, where these extra levels of functionality can be included in the prototype. For larger types of systems development there may be as many as five of these different levels (as described earlier).

Obviously, in small systems some of these levels can be grouped together, and in some particular instances all the factors relating to a later level may have been identified intuitively and included in an earlier version. The fact remains that the approach provides a vehicle for the rigorous estimation, monitoring and control of an evolutionary development process.

This is clearly not the only way in which evolutionary design components can be separated and classified. However, unless some such approach is taken, then the project control benefits which should be inherited from the structured systems methodologies of the last decade will not be realised.

1.6   Early Implementation

In an evolutionary development approach, all opportunities for early delivery of all or part of the system must be explored. For example, the possible 'phasing' of the system delivery should be considered when the first discussions on the project are taking place. It may be that some parts of the overall system are almost integral, and can be implemented earlier than the rest of the project, with minimum overheads incurred. Alternatively, perhaps in a system to be introduced into a number of similar locations, some of the locations only require part of the full system; again this sub-set of system functions could be completed and delivered early.

There is also another way in which the early delivery of a system can be brought about using an evolutionary development method. In some special circumstances, an early prototype with somewhat limited functionality can be released for use as a real system, and the later 'versions' can be treated as groups of amendments which may be issued as new releases. An example of the kind of situation in which this approach might be viable is one where there is no existing system, and where any information from a reduced form of the required system is better than no information at all. Nevertheless, such an approach should be treated with extreme caution.

1.7   Flexibility and Scalability

An evolutionary development methodology must be able to handle the development of a variety of different types of systems, some heavily data-oriented, some with very complex processing, some with critical response time problems, others with rigorous control requirements, etc. It must also be able to tackle systems of different sizes, ranging from the relatively trivial two-person-day project through to the several-person-year major development.

This means that the various tools and techniques used within the methodology must have several levels of complexity, and the analyst is required to make decisions as to the level needed for each particular case. It must also be possible to leave out some of the stages in the full methodology framework when the system to be developed is small and relatively straightforward.

One of the key requirements of an evolutionary methodology is that it can be easily tailored:

1. It may need to be adjusted to suit the type of organisation or department in which it is to be used.

2. There will almost certainly have to be slightly different versions of the methodology for each of the 4GL environments of which the organisation is making use.

3. It must be adapted to suit each of the systems development projects to which it is to be applied.

2 THE SHORT AND LONG-TERM FUTURE

The future is unpredictable; perhaps the only thing certain about it is that things will always change. Changes will occur not only in the way we carry out our business, but also in the way in which we develop our business systems. Systems Development departments within organisations must be ready to respond to these changes, and in order to be able to do that, they must be using an overall approach which is flexible enough to adapt to the new methods, with the minimum amount of upheaval.

2.1 Computer Assisted Systems Engineering

One of the most important changes affecting structured systems methodologies during the last few years has been the gradual development and use of application development support packages, such as Excelerator from ITC and Auto-mate from LBMS, which not only provide graphical facilities for creating and adjusting the different types of diagram for specific methodologies, but also automatically incorporate a high level of validation and cross-checking. These packages are one aspect of a much broader trend in the use of software to support the whole process of systems development. The software systems involved are known as Computer Aided Software Engineering (CASE) Tools, and can be considered to include the kind of fourth generation environment software discussed in chapter 1, as well as packages to support strategy planning and project management.

The big advantage that CASE tools provide is that many of the tedious cross-checking steps in a particular methodology are automated, thereby enabling the delivery of a faster and more accurate system specification. It is assumed that eventually all but the most creative components in the analysis and design process will be dealt with in this way. However, while these tools can be seen to minimise some of the disadvantages of the cookbook approach, there can still be problems in that they tend to support the formality rather than the flexibility of the framework, making it difficult to take a toolbox approach.

There are in fact now a number of integrated groups of CASE tools attempting to cover the full systems development life cycle. These groups are known as IPSEs (Integrated Project Support Environments), and are at a relatively early stage in their evolution. However they clearly have a major part to play in future systems development: There is a well known saying in Britain, 'the cobbler's children are always the last to be shod': of all the various design and engineering disciplines, we in the computer systems design business seem to be the last to make use of CAD/CAM!

It may appear surprising that, as structured systems analysis, fourth generation development and CASE are so obviously inextricably linked, so little mention of CASE has been made throughout the earlier parts of the book. There are two reasons for this. The first is that CASE technology is still in a very early stage of development, and is therefore relatively unstable; improvements and advances are occurring regularly, and rendering earlier ideas obsolete. As a result, it is quite difficult to formalise its relationship with the other two topics. The second reason is that CASE is simply the process of automating the task of systems development; it is imperative that the systems development process is completely understood and got right BEFORE attempts are made to automate it! Our concern in this book has been with that development process.

2.2   Fifth Generation Concepts

Enthusiasm for expert and knowledge-based systems ideas has waxed and waned throughout the last twenty years. At last however there seems to be some evidence of progress; a number of business systems have been developed successfully using these fifth generation concepts, and are proving the worth of the approach.

It is not considered that such methods can ever fully replace those of the third and fourth generations, but it is expected that at some time in the future the two contrasting approaches will somehow be reconciled. Bearing in mind the fact that knowledge-based systems, almost by definition, involve an evolutionary approach to knowledge acquisition, the move outlined in this book from pre-specification to evolutionary development can be seen as an important step towards that integration.

CASE and Fifth Generation are only two of the more obvious areas of change occurring within the systems development field. There are also important shifts taking place within structured systems analysis, such as the gradual swing from function and data-oriented approaches towards the use of object-oriented concepts. In fact, never before in the history of system development has there been such an abundance of new ideas in circulation, all complementing or competing with each other!

So, what is the poor systems development manager to do? There is a strong temptation to assume that eventually the situation will settle down, and industry standards will evolve. Some managers feel that it is better to wait until this happens before making any real changes to their (traditional) systems development approach. However, this means that they will miss many opportunities of saving costs and producing better systems. The alternative is to embrace change regularly, but to choose direction carefully, using judgement as to the short and long term benefits.

3   CONCLUSION

This book has been designed to make the case for an evolutionary approach to the development of business computer systems. During the last sixteen chapters we have examined in some detail all the most important aspects of the approach, its meaning, its detail, its history, its justification, its niche, its potential, and perhaps most importantly, the implications it embraces for both user and IT professional.

One of the most significant of these implications is that the evolutionary systems development approach puts back into the hands of the analyst and designer, decisions regarding the method of design to be used. This demands a higher level of skill, and a different kind of skill from that required of an analyst working as a team member using a 'cookbook' methodology. It places a heavier burden of responsibility on both the analyst and the user, but it can repay this extra commitment, with interest, by providing major benefits in terms of a shorter development time, reduced development costs, and a system which is much more user-acceptable.

The philosophy is different from that used in pre-specified systems analysis and design:

> The **PRE-SPECIFICATION** philosophy involves a view
> of the development as a 'one-off' construction, and it sees
> virtue in the pursuit of maximum safety in design, enforced

stages and methods, and an acceptance of increased overheads to minimise project risk.

The **EVOLUTIONARY** approach espouses the maximum user involvement in design, maximum use of analyst's skill to reduce overheads, and earliest possible delivery. Project risk is considered to be minimised by the heavy user involvement and responsibility from an early stage in the project. The view of development here is that one should be able to 'explore' a solution before committing to it.

The problem inherent in the pre-specification, structured cookbook methodology approach is that every aspect of the development is formalised, and all the steps and stages are required to be carried out in a specific way. This is seen as an 'engineering' approach.

However, the DESIGN of anything is a highly creative activity, and the placing of formal restrictions and limitations on how the design is to be done can seriously impede this creative process. In other engineering disciplines it is the 'construction' phase which is formalised and controlled, not the original development of an idea. An architect or designer would not consider subjugating their creative talents in favour of a slavish obedience to formalised design.

On the other hand, in an evolutionary development approach, both the design and construction are taking place at the same time, so here there are problems of adhering to formalised construction techniques while making use of the essential design flexibility provided by a 'trial and error' process.

Bearing these points in mind, this book has put forward ideas and examples designed to achieve the 'best of both worlds', by amalgamating all of the most important evolutionary design concepts with the core of the well-established structured systems analysis and design techniques and controls. The successful implementation of such an approach should enable us to cope with the natural, inevitable changes to business and computer systems in a much simpler, more effective manner, through a series of progressive rather than traumatic adjustments.

At the time of writing, there seems to be an increasing acceptance of these evolutionary concepts, though, still, most organisations have not managed to reconcile them with the earlier structured systems approach. This book is offered as a contribution towards that reconciliation.

Appendix 1   CASE STUDY

In the otherwise modest Gloucestershire village of Ambleside near Cheltenham, there stands the imposing complex of buildings which constitutes the main factory and offices of the well known and respected Gentry Shoe Company. The company has been in existence for over 150 years, and the present Chairman and Managing Director, Richard Gentry, can trace back its origins to a cobblers shop in nearby Chadwell, run by his great-great grandfather.

The company's product range includes all the best in traditional styles, all made up using best quality leather; (no substitute materials have ever been considered, although some of the more modern styles do have rubber soles). For instance, the catalogue includes a full range of stout country shoes for both ladies and gentlemen, as well as a broad selection of dress shoe styles. A much smaller selection of boots, including several varieties of riding boot, are offered.

Although Gentry are clearly very traditional in their approach to fashion, materials and market sector, they are recognised to be well up to date in their approach to management and production. They were one of the first firms to make use of computers, and currently have a thriving DP Department. Not only are the standard administrative systems of accounts, payroll, stock control and sales forecasting heavily computerised, but the production planning and scheduling are also supported by computer systems. Recently, members of the Research and Development Department went over to the United States to examine advances in computer aided design in the shoe industry, and as a result of this a state of the art CAD/CAM workstation, the first of its kind in Britain, has just been installed. Great things are hoped for from it.

There are two main problems currently facing the DP Department. The first is that most of the day-to-day operational systems, such as Sales Order Processing, Stock Control and Materials Purchasing, were originally designed some time ago, and were constructed in a mixture of third generation language and assembly code. As a result, they are now extremely difficult to amend (particularly those parts written in assembly code, as there are only two programmers left with any experience of pre-3GL programming).

The second problem is that the large mainframe computer which is used for running all these operational systems is coming to the end of its natural life. The manufacturers of the machine are unlikely to provide maintenance and support beyond the end of the existing contract, which expires in eighteen months time.

A decision has been made that the three major systems just mentioned will be analysed and re-designed over the next year, and that a modern mini-computer network will be used, together with a full fourth generation environment including an advanced database management system. The company expects that the new systems will provide a much greater level of flexibility, in the form of on-line access and update, and it is recognised that this will cause major changes to existing manual and clerical procedures.

The current Sales Order Processing system handles the receipt of something like 500 orders per week, from Gentry's own shops, and from other outlets, mainly Department Stores. Each order contains a substantial number of items. Orders are batched and checked, and fed into a Data Preparation section for entry into the computer system. The Order Processing programs are run regularly several times per day, and produce Picking Lists and error messages....

The Stock Control system supports the storage and maintenance of both the materials, which are used in the production of the shoes, and the finished products, the shoes themselves. Information is held on current stock levels, and this is updated daily (in batch mode) from sales delivery transactions, Goods In details, and production requisitions.... The computer system also supports the rolling stock-taking procedures carried out within the organisation.

The Materials Purchasing system is again a batch-oriented system, which is run overnight. It creates purchase orders (which are dispatched the following morning) and provides information on supplier deliveries. The purchasing process works on a 'call-off' system, whereby contracts are negotiated and agreed with suppliers in advance, and a purchase order simply informs the supplier that some quantity of the contracted materials is required to be delivered. It is hoped that any new Purchasing system will provide additional support for this negotiation process.

The various examples of stuctured analysis and design models shown in this book represent excerpts from the working papers of the project team responsible for carrying out this integrated development.

Appendix 2   BIBLIOGRAPHY

1   BOOKS

C. Ashworth & M Goodland, *SSADM A Practical Approach*, McGraw-Hill 1990

Bernard H. Boar, *Applications Prototyping*, Wiley-Interscience 1984

B.W. Boehm, *Software Engineering Economics*, Prentice Hall 1981

Tom DeMarco, *Structured Analysis and Systems Specification*, Yourdon Press 1979

Tom DeMarco, *Controlling Software Projects*, Yourdon Press 1982

C. Gane, *Rapid Systems Development*, Prentice Hall 1990

Tom Gilb, *Principles of Software Engineering Management*, Addison-Wesley 1988

Kit Grindley, *4GLs, A Survey of Best Practice* , IDPM Publication 1986

Charles Handy, *The Age of Unreason*, Business Books Ltd. 1989

D.C. Ince & S. Hekmatpour, *Software Prototyping in the Eighties*, Information Technology Briefing   Open University   1986

D. Law , *Prototyping: A State-of-the-art Report* ,   NCC 1985

Kenneth E. Lantz, *The Prototyping Methodology*, Prentice Hall 1984

M.M. Lehman & L.A. Belady, *Program Evolution*, Academic Press 1985

James Martin, *An Information Systems Manifesto* , Prentice Hall 1984

James Martin, *Applications Development without Programmers*, Prentice Hall 1985

James Martin, *Fourth Generation Languages*, Prentice Hall 1985

P.J. Mayhew, *Investigation of Information Systems Prototyping*, PhD Thesis   University of East Anglia   1987

E. Mumford, *Designing Participatively,* Manchester Business School 1983

J. Neilsen, *Hypertext and Hypermedia,* Academic Press 1990

T.W. Olle et al., *Information Systems Methodologies, A Framework for Understanding,* Addison-Wesley 1988

A. Sutcliffe, *Jackson Structured Development,* Prentice Hall 1989

B.J. Travis, *Auditing the Development of Computer Systems,* Butterworths 1987

R. Veryard, *Pragmatic Data Analysis,* Blackwell Scientific Publications 1984

R. Vonk, *Prototyping; The effective use of CASE technology,* Prentice Hall 1990

Gerald M. Weinberg, *The Psychology of Computer Programming,* Van Nostrand Reinhold 1971

E. Yourdon, *Structured Walkthroughs,* Yourdon Press 1978

2   ARTICLES

A.J. Albrecht, "Measuring Application Development Productivity", Proceedings of SHARE/GUIDE/IBM Symposium October 1979

P.A. Dearnley & P.J Mayhew, "In favour of Systems Prototypes", *BCS Computer Journal* Vol 26 1983

P.J. Mayhew & P.A. Dearnley, "Organisation and Management of Systems Prototyping", *Information & Software Technology* May 1990

C. Floyd, "A Systematic Look at Prototyping" from *Approaches to Prototyping ,* Eds Budde et al. Springer-Verlag 1984

D.D. McCracken & M.A. Jackson, "Lifecycle concept considered harmful" ACM SIGSOFT Software Engineering Notes 1982

A. Milton Jenkins, "Prototyping: A Methodology for the Design and Development of Application Systems", *Spectrum Magazine,* Society Information Management, Chicago April 1985

J.D. Naumann & A. Milton Jenkins,   "Prototyping: The New Paradigm for Systems Development" ACM

G. Rzevski, "Prototypes versus Pilot Systems: Strategies for Evolutionary Information System Development", from   *Approaches to Prototyping ,* Eds Budde et al.   Springer-Verlag 1984

3   MANUALS

Business Systems Planning, Information Systems Planning Guide, 2nd ed. IBM Corp., White Plains, NY 1978

Systemscraft Manual,   vols 1 & 2,   The City University, London, (internal) 1989

INDEX